Scientific Terminology for Agricultural Biotechnology

Neeraj Pratap Singh
M.Sc.(Horticulture), MBA
5/839, Viram Khand, Gomti Nagar
Lucknow (UP) India

International Book Distributing Co.
(Publishing Division)

Published by

INTERNATIONAL BOOK DISTRIBUTING CO.
(Publishing Division)
Khushnuma Complex Basement
7, Meerabai Marg (Behind Jawahar Bhawan)
Lucknow 226 001 U.P. (INDIA)
Tel. : 91-522-2209542, 2209543, 2209544, 2209545
Fax : 0522-4045308, E-Mail : ibdco@airtelmail.in

Price: Rs. 225/-

ISBN 81-8189-280-1
First Edition 2009

Composed & Designed at :
Panacea Computers
3rd Floor, Agarwal Sabha Bhawan, Subhash Mohal
Sadar Cantt. Lucknow-226 002, Phone : 0522-2483312, 9335927082
E-mail : prasgupt@rediffmail.com; prasgupt@hotmail.com

Printed at:
Salasar Imaging Systems
C-7/5, Lawrence Road Industrial Area
Delhi - 110 035
Tel. : 011-27185653, 9810064311

तेरा तुझको अर्पण

Preface

Biotechnology is a general term used in a very broad field of study. According to the Convention on Biological Diversity, biotechnology means:

"Any technological application that uses biological systems, living organisms, or derivatives thereof, to make or modify products or processes for specific use."

Interpreted in this broad sense, the definition covers many of the tools and techniques that are common place today in agriculture and food production. If interpreted in a narrow sense, the definition considers only the "new" DNA, molecular biology and reproductive technology.

The swiftness of change in the sector means that terminology is constantly evolving and yesterday's buzzword is today's jargon and might be tomorrow's mainstream term. The rate of evolution of terminology has been such that it has been very difficult to remain abreast of current usages.

The idea for such a collection of terminology associated with the rapidly expanding fields related to or deriving from biotechnology and genetic engineering, was stimulated by the difficulty of communicating effectively in discussions at intergovernmental level. On various occasions, simple differences of interpretation of terminology have threatened to de-rail negotiations of international importance.

It tries to provide a consolidated, comprehensive and yet accessible list of terms and acronyms that are used regularly in biotechnology *sensu lato* and in the very broad area commonly dubbed "genetic engineering," with all the associated problems of usage of originally discrete technical terms in a general context by a mass media that does not discriminate, or in a legal context that requires very exact definitions.

This Scientific Terminology is an attempt to present an up-to-date list of terms currently in use in biotechnology, genetic engineering and closely allied fields. It is intended to provide a convenient reference source for researchers, students and technicians.

The Glossary has been prepared in response to an expressed need. Many of the terms listed in this volume are otherwise found only in published papers and books. The terms included have been selected by examination of books, dictionaries, journals and abstracts dealing entirely or in part with biotechnology or allied fields.

In addition, an attempt has been made to include terms from applied biotechnology that are important for intergovernmental activities and especially in the areas of plant and animal genetic resources, food quality and plant protection.

Neeraj Pratap Singh

Contents

A

A Abbreviation for **adenine**.

Ab Abbreviation for **antibody**.

ABC model Widely accepted model of flower organ identity that appears generally applicable to distantly related **dicotyledon**ous, although less well to **monocotyledon**ous plants. The model incorporates the *Arabidopsis* genes required for flower organ identity.

abiotic Absence of living organisms.

abscisic acid A **phytohormone** implicated in the control of many plant responses to **abiotic** stress, such as extent of stomatal opening under water deficit (i.e. drought) conditions.

abzyme *See*: **catalytic antibody**.

acaricide A **pesticide** used to kill or control mites or ticks.

ACC synthase Abbreviation for 1-aminocyclopropane-1-carboxylase. The **enzyme** catalyses the rate limiting step in the **ethylene** biosynthetic pathway, and is particularly significant in the fruit ripening process. Plants typically carry a number of distinct ACC synthase genes, which are differentially regulated in response to a variety of developmental, environmental and chemical factors.

acceptor control The regulation of the rate of respiration by the availability of **ADP** as a phosphate acceptor.

acceptor junction site The junction between the 3' end of an **intron** and the 5' end of an **exon**. *See*: **donor junction site.**

accessory bud A **lateral bud** occurring at the base of a **terminal bud** or at the side of an **axillary bud.**

acclimatization The **adaptation** of a living **organism** (plant, animal or microorganism) to a changed environment that subjects it to physiological stress. Acclimatization should not be confused with **adaptation.**

acellular Tissues or organisms that are not made up of separate cells but often have more than one nucleus.

acentric chromosome Chromosome fragment lacking a **centromere**.

acetyl CoA Abbreviation for **acetyl co-enzyme A.**

acetyl co-enzyme A (Abbreviation: acetyl CoA) A compound formed in the mitochondria when an acetyl group (CH_3CO^-) - derived from breakdown of fats, **proteins**, or carbohydrates - combines with the thiol group (-SH) of **co-enzyme** A.

ACP Abbreviation for **acyl carrier protein**.

acquired Developed in response to the environment, not inherited, such as a character trait (acquired characteristic) resulting from environmental effect(s). *cf* **acclimatization.**

acridine dyes A class of positively charged polycyclic molecules that intercalate into **DNA** and induce frameshift mutations.

acrocentric A **chromosome** that has its **centromere** near the end.

acropetal Arising or developing in a longitudinal sequence beginning at the base and proceeding towards the apex. *Opposite*: **basipetal**.

activated carbon *See* **activated charcoal.**

activated charcoal Charcoal that has been treated to remove hydrocarbons and to increase its adsorptive properties. It acts by condensing and holding a gas or solute onto its surface; thus inhibitory substances in nutrient medium may be adsorbed to charcoal included in the medium.

active transport The movement of a molecule or groups of molecules across a **cell** membrane, which requires the expenditure of cellular energy, because the direction of movement is against the prevailing concentration gradient.

acute transfection Short-term **transfection**.

acyl carrier protein (Abbreviation: ACP). A class of molecules that bind acyl intermediates during the formation of long-chain fatty acids. ACPs are important because of their involvement in many of the reactions necessary for *in vivo* fatty acid synthesis.

adaptation Adjustment of a **population** to changes in environment over generations, associated (at least in part) with genetic changes resulting from **selection** imposed by the changed environment. *Not* **acclimatization.**

additive genes Genes whose net effect is the sum of their individual **allelic** effects, i.e. they show neither **dominance** nor **epistasis**.

additive genetic variance The net effect of the expression of additive genes, and thus the chief cause of the resemblance between relatives. It represents the main determinant of the response of a **population** to selection. Formally, the **variance** of **breeding values**.

adenine (Abbreviation: A). One the **bases** found in **DNA** and **RNA**. *See*: **adenosine**.

adenosine diphosphate (adenosine 5'-diphosphate) (Abbreviation: ADP). *See*: **adenosine triphosphate**.

adenosine monophosphate (adenosine 5'-monophosphate) (Abbreviation: AMP). *See*: **adenylic acid, adenosine triphosphate**.

adenosine The (ribo)**nucleoside** resulting from the combination of the base **adenine** (A) and the sugar D-**ribose**. The corresponding

deoxyribonucleoside is called deoxy-adenosine. *See*: **adenosine triphosphate, adenylic acid, dATP**.

adenosine triphosphate (adenosine 5'-triphosphate) (Abbreviation: ATP). A **nucleotide** of fundamental importance as the major carrier of chemical energy in all living organisms. It is also required for **RNA** synthesis, since it is a direct precursor molecule. ATP consists of **adenosine** with three phosphate groups, linked together linearly. The phosphates are attached to adenosine through the 5'-hydroxyl of its **ribose** (sugar) portion. Upon **hydrolysis**, these bonds yield either one molecule of adenosine 5'-diphosphate (ADP) and the inorganic phosphate ion, or one molecule of **adenosine 5'-monophosphate** (AMP) and **pyrophosphate**; in both cases releasing energy that is used to power biological processes. **ATP** is regenerated by the **phosphorylation** of AMP and ADP.

adenovirus One of a group of **DNA**-containing viruses found in rodents, fowl, cattle, monkeys, and man. In man, they are responsible for respiratory-tract infections, but they have been exploited as a **vector** in **gene therapy**, especially for genes targeted at the lungs.

adenylic acid Synonym for **adenosine monophosphate**, a (ribo)**nucleotide** containing the **nucleoside** adenosine. The corresponding **deoxyribonucleotide** is called **deoxyadenosine** 5'-monophosphate or deoxyadenylic acid.

adoptive immunization The transfer of an immune state from one animal to another by means of **lymphocyte** transfusions.

ADP Abbreviation for **adenosine diphosphate**.

adventitious A structure arising at sites other than the usual ones, e.g. shoots from roots or leaves, and embryos from any cell other than a **zygote**.

aerobe A **microorganism** that grows in the presence of oxygen. *Opposite*: **anaerobe**.

aerobic Active in the presence of free oxygen, e.g. aerobic bacteria that can live in the presence of oxygen.

aerobic respiration A type of respiration in which foodstuffs are completely oxidized to carbon dioxide and water, with the release of chemical energy, in a process requiring atmospheric oxygen.

affinity chromatography A method for purifying specific components in a solution by exploiting their specific binding to known molecule(s). The mixed solution is passed through a column containing a solid medium to which the binding molecule is covalently attached. *See*: **immunoaffinity chromatography; metal affinity chromatography; pseudo-affinity chromatography**.

affinity tag An amino acid sequence that has been engineered into a **protein** to make its purification easier. The tag could be another **protein** or a short **amino acid** sequence, allowing purification by **affinity chromatography**. *Synonym*: **purification tag**.

aflatoxins A group of toxic compounds, produced by *Aspergillus flavus*, that bind to **DNA** and prevent **replication** and transcription. Aflatoxins can cause acute liver damage and cancer. A health hazard in certain stored foods or feed.

AFLP Abbreviation for **amplified fragment length polymorphism**.

Ag Abbreviation for **antigen**.

agar A **polysaccharide** gelifying agent used in nutrient media preparations and obtained from *Rhodophyta* (red algae). Both the type of agar and its concentration can affect the growth and appearance of cultured **explants**.

agarose gel electrophoresis A method to separate **DNA** and **RNA** molecules on the basis of their size, in which samples are subjected to an electric field applied to a gel made with **agarose**.

agarose The main functional constituent of **agar**.

aggregate 1. A clump or mass formed by gathering or collecting units. 2. A body of loosely associated cells, such as a friable callus or cell suspension. 3. Coarse inert material, such as gravel, that is mixed with soil to increase its porosity. 4. A serological reaction in which the **antibody** and **antigen** react and precipitate.

agonist A drug, **hormone** or transmitter substance that forms a complex with a **receptor** site. The formation of the complex triggers an active response from a cell.

Agrobacterium A genus of bacteria that includes several plant pathogenic species, causing tumour-like symptoms. *See*: ***Agrobacterium rhizogenes, Agrobacterium tumefaciens.***

Agrobacterium rhizogenes A bacterium that causes **hairy root** disease in some plants. Similar to the **crown gall** disease caused by ***Agrobacterium tumefaciens,*** this is achieved by the mobilization of the bacterial **Ri plasmid** with the transfer to the plant of some of the genetic material from the plasmid. This process has been used to insert foreign genes into plant cells, but to a lesser extent than the ***Agrobacterium tumefaciens*-mediated transformation** system, because **regeneration** of whole plants from **hairy root cultures** is problematical.

Agrobacterium tumefaciens A bacterium that causes **crown gall** disease in some plants. The bacterium characteristically infects a wound, and incorporates a segment of **Ti plasmid DNA** into the host **genome**. This **DNA** causes the host cell to grow into a tumour-like structure that synthesizes specific **opines** that only the pathogen can metabolize. This **DNA**-transfer mechanism is exploited in the genetic engineering of plants. *See*: **T-DNA.**

Agrobacterium tumefaciens*-mediated transformation** The process of **DNA** transfer from ***Agrobacterium tumefaciens to plants, that occurs naturally during **crown gall** disease, and can be used as a method of **transformation**.

AHG Abbreviation for **antihaemophilic globulin.**

AI Abbreviation for **artificial insemination**.

airlift fermenter A cylindrical **fermentation** vessel in which the cells are mixed by air introduced at the base of the vessel and that rises through the column of culture medium. The

cell suspension circulates around the column as a consequence of the gradient of air bubbles in different parts of the reactor.

albinism Hereditary absence of **pigment** in an organism. **Albino** animals have no colour in their skin, hair and eyes. Albino plants lack chlorophyll.

albino 1. An organism lacking pigmentation, due to genetic factors. The condition is **albinism** 2. A conspicuous **plastid** mutant involving loss of chlorophyll.

aleurone The outermost layer of the **endosperm** in a seed, and the site of enzymes concerned with endosperm digestion during seedling growth.

algal biomass Single-celled plants (e.g. *Chlorella* spp. and *Spirulina* spp.) grown commercially in ponds to make feed materials for zooplankton, which are in turn harvested as feed for fish farms.

alginate Polysaccharide gelling agent.

alkylating agent A class of chemicals that transfer alkyl (methyl, ethyl, etc.) groups; for example to the bases in **DNA**. Some of these (especially ethyl methane sulphonate, abbreviated EMS) have been much used as **mutagens**.

allele A variant form of a **gene**. In a **diploid** cell, there are two alleles of every gene (one inherited from each parent, although they could be identical). Within a **population,** there may be many alleles of a gene. Alleles are symbolized with a capital letter to denote **dominance**, and lower case for **recessive**. In **heterozygotes** with co-dominant alleles, both are expressed. *See*: **multiple alleles**. *Synonym*: allelomorph.

allele frequency The relative number of copies of an allele in a population, expressed as a proportion of the total number of copies of all alleles at a given locus in a population.

allele-specific amplification (Abbreviation: ASA). The use of the **polymerase chain reaction** at a sufficiently high **stringency** that only one **allele** is amplified. A powerful means of genotyping for single-locus disorders that have been characterized at the molecular level.

allelic (adj.) *See* **allele.**

allelic exclusion A phenomenon whereby only one functional **allele** of an **antibody** gene can be assembled in a given B **lymphocyte**.

allelomorph *See*: **allele.**

allelopathy The **secretion** of chemicals, such as phenolic and terpenoid compounds, by a plant's roots, which inhibit the growth or reproduction of competitor plants.

allergen An **antigen** that provokes an **immune response.**

allogamy Cross fertilization in plants. *See*: **fertilization.**

allogenic Differing at one or more loci, although belonging to the same species. Thus an organ transplant from one human donor to another is allogeneic, whereas a transplant from a baboon to a human would be **xenogeneic**.

allometric When the **growth rate** of one part of an organism differs from that of another part or of the rest of the body.

allopatric In the context of natural populations of animals or plants, inhabiting distinct and separate areas.

allopatric speciation Speciation occurring at least in part because of geographic isolation.

allopolyploid A **polyploid** organism with sets of chromosomes derived from different species. *Opposite*: **autopolyploid**.

allosome Synonym for **sex chromosome**.

allosteric control *See*: **allosteric regulation**.

allosteric enzyme An enzyme that has two structurally distinct forms, one of which is active and the other inactive. Active forms tend to catalyse the initial step in a pathway leading to the synthesis of molecules. The end product of this synthesis can act as a feedback inhibitor, converting the enzyme to the inactive form, thus controlling the amount of product synthesized. *Synonym*: **allozyme**.

allosteric regulation A catalysis-regulating process in which the binding of a small **effector molecule** to one site on an **enzyme** affects the activity at another site.

allosteric site That part of an **enzyme** molecule

where the non-covalent binding of an **effector molecule** can affect the enzyme's catalytic activity. *See*: **conformation, ligand.**

allosteric transition A reversible interaction of a small molecule with a **protein** molecule, resulting in a change in the shape of the **protein** and consequent alteration of the interaction of that **protein** with a third molecule.

allotetraploid An **allopolyploid** having two different progenitor genomes.

allotype A classification of **antibody** molecules according to the antigenicity of the constant regions; a variation that is determined by a single allele.

allozygote An individual that is heterozygous for two different mutant **alleles**.

allozyme *See*: **allosteric enzyme.**

alpha globulin *See*: **haptoglobin.**

alternative mRNA splicing The inclusion or exclusion of different **exons** to form different **mRNA transcripts** from a single **transcription** unit.

Alu sequences A highly repeated family of 300-bp long sequences dispersed throughout the human genome, so named because they are released by the digestion of genomic **DNA** with the **restriction endonuclease** AluI.

amber stop codon *See*: **stop codon.**

amino acid A compound containing both amino ($-NH_2$) and carboxyl (-COOH) groups. In particular, any of 20 basic building blocks of **proteins** having the formula NH_2-CR-COOH, where R is different for each specific amino acid. *See*: annex 3.

aminoacyl site (Abbreviation: A-site). One of two sites on **ribosomes** to which the aminoacyl **tRNA** molecules can bind.

aminoacyl tRNA synthetase An **enzyme** that catalyses the attachment of an **amino acid** to its specific **tRNA** molecule.

amitosis A **cell division** (including nuclear division through constriction of the nucleus) that occurs without **chromosome** differentiation as in **mitosis**. The mechanism whereby the genetic integrity is maintained during amitosis is uncertain.

amniocentesis A procedure for obtaining foetal cells for prenatal diagnosis by sampling the **amniotic fluid** from a pregnant mammal. Cells are cultured, and the **karyotype** is checked for known irregularities (e.g. Down's syndrome and spina bifida in humans).

amnion The thin membrane that lines the fluid-filled sac in which the **embryo** develops in higher vertebrates, reptiles and birds.

amniotic fluid Liquid contents of the amniotic sac of higher vertebrates, containing foetal, but not maternal cells.

amorph A mutation that abolishes gene function. *Synonym*: **null mutation**.

AMP Abbreviation for **adenosine monophosphate.**

amphidiploid A plant derived from doubling the **chromosome** number of an interspecific F_1 hybrid. Naturally found hybrids of this sort are referred to as **allopolyploid.**

amphimixis True sexual **reproduction** involving the fusion of male and female gametes and the formation of a **zygote**.

ampicillin A penicillin-type **antibiotic** that prevents bacterial growth by interfering with synthesis of the cell wall. Commonly used as a **selectable marker** in the creation of **transgenic** plants.

amplicon The product of a **DNA amplification** reaction. *See*: **polymerase chain reaction.**

amplification 1. Creation of many copies of a segment of **DNA** by the **polymerase chain reaction.** 2. Treatment (e.g. use of chloramphenicol) designed to increase the proportion of **plasmid DNA** relative to that of bacterial (host) **DNA**. 3. Evolutionary expansion in copy number of a **repetitive DNA** sequence through a process of repeated duplication.

amplified fragment length polymorphism (Abbreviation: AFLP). A type of **DNA marker,** generated by the **PCR** amplification of **restriction endonuclease** treated **DNA**. A small proportion of all **restriction fragments** is amplified in any one reaction, so that AFLP profiles can be analysed by gel electrophoresis. This has the important characteristic that many markers can be generated with relatively little effort.

amplify To increase the number of copies of a **DNA** sequence, either *in vivo* by inserting into a **cloning vector** that replicates within a host cell, or *in vitro* by **polymerase chain reaction**.

ampometric *See:* **electrochemical sensor**

amylase Describing a wide class of enzymes that catalyse the **hydrolysis** of starch.

amylolytic The capability of enzymatically degrading **starch** into sugars.

amylopectin A **polysaccharide** comprising highly branched chains of glucose **residues**. The water-insoluble portion of **starch**.

amylose A **polysaccharide** consisting of linear chains of 100-1000 glucose **residues**. The water-soluble portion of **starch**.

anabolic pathway A pathway by which a **metabolite** is synthesized; a biosynthetic pathway.

anabolism One of the two subcategories of **metabolism**, referring to the building up of complex organic molecules from simpler precursors.

anaerobe An organism that can grow in the absence of oxygen. *Opposite*: **aerobe**.

anaerobic An environment or condition in which molecular oxygen is not available for chemical, physical or biological processes.

anaerobic digestion Digestion of materials in the absence of oxygen. *See*: **anaerobic respiration**.

anaerobic respiration Respiration in which foodstuffs are partially oxidized, with the release of chemical energy, in a process not involving atmospheric oxygen. A notable example is in alcoholic fermentation, where sugar is metabolized into ethanol.

analogous Features of organisms or molecules that are superficially or functionally similar but have evolved in a different way or contain different compounds.

anaphase The stage of **mitosis** or **meiosis** during which the daughter chromosomes migrate to opposite poles of the cell (toward the ends of the **spindle**). Anaphase follows **metaphase**

and precedes **telophase**.

anchor gene A gene that has been positioned on both the **physical map** and the **linkage map** of a chromosome, and thereby allows their mutual alignment.

androgen Any hormone that stimulates the development of male secondary sexual characteristics, and contributes to the control of sexual activity in vertebrate animals. Usually synthesized in the **testis**.

androgenesis Male **parthenogenesis**, i.e. the development of a haploid embryo from a male nucleus. The maternal nucleus is eliminated or inactivated subsequent to **fertilization** of the ovum, and the **haploid** individual (referred to as androgenetic) contains in its cells the genome of the male **gamete** only. *See*: **anther culture**; **gynogenesis**.

aneuploid An organism or cell having a chromosome number other than the normal **somatic** number. Aneuploid gametes have a **chromosome** number other than the normal **haploid** number. The condition is *aneuploidy*.

angiogenesis The formation and development of new blood vessels in the body, stimulated by growth factors, such as **angiogenin**. The process is required for the spread of malignant tumours.

angiogenin One of the human angiogenic growth factors. In addition to stimulating (normal) blood vessel formation, angiogenin levels are correlated with placenta formation and tumour growth.

angiosperm A division of the plant kingdom that includes all flowering plants, i.e. **vascular** plants in which double fertilization occurs resulting in development of fruit containing seeds. Divided into two major groups, **monocotyledons** and **dicotyledons**. *See:* gymnosperm.

animal cell immobilization Entrapment of animal cells in some solid material in order to produce some natural product or genetically engineered **protein**. Animal cells have the advantage that they already produce many **proteins** of pharmacological interest, and that genetically engineered **proteins** are produced by them with the post-translation modifications normal to animals. However, because animal cells are much more fragile than bacterial ones, they cannot tolerate a commercial **fermentation** process.

animal cloning *See*: **cloning**.

anneal The pairing of **complementary DNA** or **RNA** sequences, via hydrogen bonding, to form a double-stranded **polynucleotide**. *Opposite*: **denature**.

annual 1. (adj:) Taking one year, or occurring at intervals of one year. 2. A plant that completes its life cycle within one year. *See* biennial, perennial.

anonymous DNA marker A **DNA marker** detectable by virtue of variation in its sequence. The function (if any) of the sequence is unknown. **Microsatellites** and **AFLPs** are typical anonymous **DNA** markers.

antagonism An interaction between two organisms (e.g. moulds or bacteria) in which the growth of one is inhibited by the other. *Opposite*: **synergism**.

antagonist A compound that inhibits the effect of an **agonist** in such a way that the combined biological effect of the two becomes smaller than the sum of their individual effects.

anther culture The aseptic culture of immature **anthers** to generate **haploid** plants from microspores via **androgenesis**.

anther The upper part of a **stamen**, containing pollen sacs within which the **pollen** develops and matures.

anthesis The period during which **anthers** bear mature and functional pollen.

anthocyanin A water-soluble blue, purple or red flavonoid pigments found in vacuoles of cells of certain plants.

antiauxin A chemical that interferes with the **auxin** response, sometimes by the prevention of auxin transport. Some antiauxins may promote **morphogenesis** *in vitro* (e.g. 2,3,5-triiodobenzoate (TIBA) and 2,4,5-trichlorophenoxyacetate (2,4,5-T)) and are therefore used to stimulate the growth of some cultures.

antibiosis The prevention of growth or **development** of an organism by a substance or another organism.

antibiotic A class of natural and synthetic compounds that inhibit the growth of, or kill some microorganisms. Antibiotics are widely used medicinally to control bacterial pathogens, but **resistance** in bacteria to particular antibiotics is often rapidly acquired through **mutation**.

antibiotic resistance marker gene (Abbreviation: ARMG). Genes (usually of bacterial origin) used as selection markers in **transgenesis**, because their presence allows cell survival in the presence of normally toxic antibiotic agents. These genes were commonly used in the development and release of first generation transgenic organisms (particularly crop plants), but are no longer favoured because of perceived risks associated with the unintentional transfer of antibiotic resistance to other organisms. *See* **kan**^{r}, *neo*^{r}.

antibiotic resistance The ability of a microorganism to disable an antibiotic or prevent its transport into the cell.

antibody (Abbreviation: Ab). An immunological **protein** produced by the **lymphocytes** in response to contact with an **antigen**. Each antibody recognizes just one **antigenic determinant** of one antigen and acts by specifically binding to it, thus rendering it harmless. Those from the IgG **antibody class** are found in the bloodstream and used in **immunoassay**. *Synonym*: immunoglobulin. *See*: **monoclonal antibody, polyclonal antibody**.

antibody binding site The part of an **antibody** that binds to the **antigenic determinant**. *See*: **complementarily-determining regions**. *Synonym*: paratope.

antibody class The class to which an antibody belongs, depending on the type of heavy chain present. In mammals, there are five classes of antibodies: IgA, IgD, IgE, IgG, and IgM.

antibody structure Describes the molecular architecture of an **antibody**, which consists of two identical "light" chains and two identical "heavy" chains and has two **antigen**-binding sites. Each chain consists of a constant region which is the same between antibodies of the same class and sub-class, and a variable region that is antibody-specific.

antibody-mediated immune response The synthesis of antibodies by B cells in response to an encounter of the cells of the immune system with a foreign **antigen**. *Synonym*: **humoral immune response**.

anticlinal The orientation of cell wall or plane of **cell division** perpendicular to the surface. *Opposite*: **periclinal**.

anticoding strand The **DNA** strand used as template for **transcription**. The resulting **mRNA** is complementary in sequence to that of the anticoding strand. *Synonym*: **template strand**.

anticodon A triplet of **tRNA** nucleotides that corresponds to a complementary **codon** in an **mRNA** molecule during **translation**.

antigen (Abbreviation: Ag). A **macromolecule** (usually a **protein** foreign to the organism), which elicits an **immune response** on first exposure to the immune system by stimulating the production of **antibodies** specific to its various **antigenic determinants**. During subsequent exposures, the antigen is bound and inactivated by these antibodies. *Synonym*: immunogen.

antigenic determinant The individual surface feature of an **antigen**, that elicits the production of a specific **antibody** in the course of the immune response. Each antigenic determinant, typically a few **amino acids** in size, causes the synthesis of a different antibody and thus exposure to a single antigen may result in the expression of a number of antibodies. *See*: **monoclonal antibody, polyclonal antibody**. *Synonym*: epitope.

antigenic switching The altering of a microorganism's surface antigens through genetic re-arrangement, to elude detection by the host's immune system.

antihaemophilic factor VIII *See*: **antihaemophilic globulin**.

antihaemophilic globulin (Abbreviation AHG). One of the blood clotting factors, a soluble **protein** that causes the fibrin matrix of a blood clot to form. Used as a treatment for haemophilia, AHG is usually obtained from genetically engineered cell cultures. *Synonym*: antihaemophilic factor VIII.

anti-idiotype antibody An **antibody**, produced by an organism, which specifically binds to the binding site of an antibody developed by that organism against a foreign **antigen**. Involved with the regulation of the immune response. Some allergic responses are in part due to the breakdown of this sort of regulation.

antimicrobial agent Any chemical or biological agent that inhibits the growth and/or survival of microorganisms. *See*: **antibiotic**.

antinutrient Compounds that inhibit the normal uptake or utilization of nutrients.

anti-oncogene A gene whose product prevents the normal growth of tissue.

antioxidant Compounds that slow the rate of oxidation reactions.

antiparallel orientation The normal arrangement of the two strands of a **double-stranded DNA** molecule, and of other nucleic-acid duplexes (**DNA-RNA, RNA-RNA**), in which the two strands are oriented in opposite directions so that the 5'-phosphate end of one strand is aligned with the 3'-hydroxyl end of the complementary strand.

antisense DNA One of the two strands of double-stranded **DNA,** usually that which is **complementary** (hence "anti") to the **mRNA**, i.e. the non-transcribed strand. However, there is not universal agreement on this convention, and the preferred designations are **coding strand** for the strand whose sequence matches that of the **mRNA**, and **non-coding strand** or template strand for the complementary strand (i.e. the transcription template).

antisense gene A gene that produces an **mRNA** complementary to the transcript of a normal gene (usually constructed by inverting the coding region relative to the **promoter**).

antisense RNA A **RNA** sequence that is complementary to all or part of a functional m**RNA** molecule, to which it binds, blocking its translation.

antisense therapy The *in vivo* treatment of a **genetic disease** by blocking translation of a **protein** with a **DNA** or an **RNA** sequence that is complementary to a specific m**RNA**.

antiseptic Any substance that kills or inhibits the growth of disease-causing **microorganism** (a microorganism capable of causing sepsis), but is essentially non-toxic to cells of the body.

antiserum The fluid portion of the blood of an immunized animal (after coagulation of the blood), which retains any **antibodies**.

anti-terminator A **protein** which enables **RNA polymerase** to ignore certain transcriptional stop or **termination signals** and thereby produce longer than normal transcripts.

antitranspirant A compound designed to reduce plant transpiration. Applied to the leaves of newly transplanted trees, shrubs etc., or cuttings in lieu of misting. Can interfere with **photosynthesis** and respiration if the coating is too thick or is unbroken.

antixenosis The modification of the behaviour of an organism by a substance or another organism. Particularly used in the context of a plant's apparent resistance against insect feeding, when the insects are presented with a choice of plant genotypes.

apex The portion of a root or shoot containing the primary or **apical meristem**.

apical cell A meristematic initial in the apical meristem of shoots or roots of plants.

apical dominance The phenomenon where growth of lateral (axillary) **buds** in a plant is inhibited by the presence of the terminal (apical) bud on the branch. Explained by the export of **auxins** from the apical bud.

apical meristem A region of the tip of each shoot and root of a plant in which cell division is continually occurring to produce new stem and root tissue, respectively. Two regions are visible in the apical **meristem**: An outer 1-4-

cell layered region (the *tunica*), where cell divisions are **anticlinal**; and below the tunica, (ii) the *corpus*, where the cells divide in all directions, and increase in volume.

apoenzyme Inactive enzyme that has to be associated with a **co-enzyme** in order to function. The apoenzyme/co-enzyme complex is called a **holoenzyme**.

apomixis The production of an **embryo** in the absence of **meiosis**. Apomictic higher plants produce **asexual** seeds, derived only from maternal tissue. *See*: **parthenogenesis**.

apoptosis The process of programmed **cell** death, which occurs naturally as a part of normal development, maintenance and renewal of tissue. Differs from **necrosis**, in which cell death is caused by external factors (stress or toxin).

AP-PCR *See*: **arbitrarily primed polymerase chain reaction**.

aptamer A polynucleotide molecule that binds to a specific molecule, often a **protein**.

aquaculture Farming of aquatic organisms, including fish, molluscs, crustaceans and aquatic plants.

Arabidopsis A genus of flowering plants in the *Cruciferae*. *A. thaliana* is used in research as a model plant because it has a small fully sequenced genome, can be cultured and transformed easily, and has a rapid generation time.

arbitrarily primed polymerase chain reaction (Abbreviation: AP-PCR) An application of the **polymerase chain reaction** to generate **DNA fingerprints**. The technique uses **arbitrary primers** to amplify anonymous stretches of **DNA**. *See*: **DNA amplification fingerprinting, random amplified polymorphic DNA**.

arbitrary primer An oligonucleotide **primer** whose sequence is chosen at random, rather than one whose sequence matches that of a known locus. These primers therefore amplify **DNA** fragments which have not been pre-selected.

Archaea Single-celled life forms adapted to existence in high pressure, anaerobic, environments such as at extreme ocean depths. These organisms are seen as a promising source of enzymes robust enough for a number of demanding industrial processes.

ARMG Abbreviation for **antibiotic resistance marker gene**.

ARS Abbreviation for autonomous(ly) replicating segment (or sequence).

artificial inembryonation Non-surgical transfer of embryo(s) to a recipient female. As *in vitro* **embryo** technology develops, artificial inembryonation may replace **artificial insemination**.

artificial insemination (Abbreviation: AI). The deposition of semen, using a syringe, at the mouth of the uterus to make conception possible.

artificial medium *See*: **culture medium**.

artificial seed Encapsulated or coated **somatic embryos** that are planted and treated like seed.

artificial selection The practice of choosing individuals from a **population** for reproduction, usually because these individuals possess one or more desirable traits.

ASA Abbreviation for **allele-specific amplification**.

ascites Abnormal accumulation of fluid in the peritoneal cavity, occurring naturally as a complication of cirrhosis of the liver, among other conditions. In the context of **monoclonal antibody** production, **hybridoma** cells are injected into mice to induce their proliferation in the resulting ascites. This method has been largely superseded by *in vitro* culture of hybridomas.

ascospore One of the spores contained in the **ascus** of certain fungi.

ascus (pl.: asci) Reproductive sac in the sexual stage of a type of fungi (*Ascomycetes*) in which ascospores are produced.

aseptic Sterile, free of contaminating organisms (bacteria, fungi, algae but not generally including viruses, and particularly not internal **symbionts**).

asexual embryogenesis *See*: **somatic cell em-**

bryogenesis.

asexual propagation Vegetative, somatic, non-sexual **reproduction** of a plant without fertilization.

asexual Reproduction not involving **meiosis** or the union of **gametes**.

asexual reproduction Reproduction that does not involve the formation and union of gametes from the different sexes or mating types. It occurs mainly in lower animals, microorganisms and plants. In plants, asexual reproduction is by vegetative propagation (e.g. bulbs, tubers, corms) and by formation of spores.

A-site Abbreviation for **aminoacyl site**.

assay 1. To test or evaluate. 2. The procedure for measuring the quantity of a given substance in a sample (chemically or by other means).

assortative mating Mating in which the partners are chosen on the basis of phenotypic similarity.

assortment *See*: segregation.

asymmetric hybrid A hybrid formed, usually via **protoplast fusion**, between two donors, where the **chromosome** complement of one of the donors is incomplete. This chromosome loss can be induced by **irradiation** or chemical treatment, or can occur naturally.

asynapsis The failure or partial failure in the pairing of homologous chromosomes during the first meiotic prophase.

ATP Abbreviation for **adenosine triphosphate**.

ATP-ase An enzyme that brings about the hydrolysis of **adenosine triphosphate**, by the cleavage of either one phosphate groups with the formation of **ADP** and inorganic phosphate, or of two phosphate groups, with the formation of **AMP** and pyrophosphate.

attenuated vaccine A virulent organism that has been modified to produce a less virulent form, but nevertheless retains the ability to elicit antibodies against the virulent form. *See*: **inactivated agent**.

attenuation A mechanism for controlling **gene expression** in prokaryotes that involves premature termination of transcription.

attenuator A nucleotide sequence in the 5' region of a prokaryotic gene (or in its **RNA**) that causes premature termination of **transcription**, possibly by forming a secondary structure.

aureofacin An antifungal **antibiotic** produced by a strain of *Streptomyces aureofaciens*. A possible candidate for the **transgenic** control of plant fungal disease.

authentic protein A recombinant **protein** that has all the properties - including any post-translational modifications - of its naturally occurring counterpart.

autocatalysis Catalysis in which one of the products of the reaction is a catalyst for the reaction.

autocatalytic reaction *See*: **autocatalysis**.

autoclave 1. An enclosed chamber in which materials can be heated under pressure to **sterilize** utensils, liquids, glassware, etc., using steam.

autogenous control The action of a gene product to inhibit (negative autogenous control) or enhance (positive autogenous control) the expression of the gene that codes for it.

auto-immune disease Disorder in which the immune systems of affected individuals produce antibodies against molecules that are normally produced by those individuals (called self **antigens**).

auto-immunity A disorder in the body's defence mechanism in which an **immune response** is elicited against its own (self) tissues.

autologous cells Cells taken from an individual, cultured (or stored), and, possibly, genetically manipulated before being transferred back into the original donor.

autolysis The process of self destruction of a **cell**, cell organelle, or tissue, through the action of lysosomic enzymes.

autonomous A term applied to any biological unit that can function on its own, i.e. without the help of another unit, such as a **transposable element** that encodes an enzyme for its own transposition.

autonomous(ly) replicating segment (or sequence) (Abbreviation: ARS). Any eukaryotic **DNA** sequence that initiates and supports chromosomal replication; they have been isolated in **yeast** cells.

autopolyploid A polyploid whose constituent genomes are derived from the same or nearly the same progenitor. In an autotetraploid, each **chromosome** is present in four copies, so meiotic configurations may include many (or exclusively) quadrivalents (four paired chromosomes), and the **inheritance** of alleles will be **quadruplex**. Quadrivalents do not always segregate normaly at meiosis, resulting in lowered fertility, so some established autotetraploid species that reproduce sexually have restricted **quadrivalent** formation.

autoradiograph A technique for visualizing the presence, location and intensity of radioactivity in histological preparations, paper chromatograms or electrophoretic gel separations, obtained by overlaying the surface with X-ray film and allowing the radiation to form an image on the film.

autosome Any of the chromosomes except the **sex chromosomes**.

autotroph Organism capable of self-nourishment utilizing carbon dioxide or carbonates as the sole source of carbon and obtaining energy from radiant energy or from the oxidation of inorganic elements, or compounds such as iron, sulphur, hydrogen, ammonium and nitrites. *Opposite*: **heterotroph.**

autotrophic (adj.) See: **autotroph.**

auxin A group of **plant growth regulators** (natural or synthetic) which stimulate cell division, enlargement, apical dominance, **root** initiation, and flowering.

auxin-cytokinin ratio The relative proportion of **auxin** to **cytokinin** present in plant **tissue culture** media. Varying the relative amounts of these two hormones affects the proportional growth of shoots and roots.

auxotroph A mutant cell or **microorganism** lacking one metabolic pathway present in the parental strain, and that consequently will not multiply on a minimal medium, but requires for growth the addition of a specific compound, such as an **amino acid** or a **vitamin**.

availability A reflection of the form and location of nutritional elements and their suitability for absorption.

avidin A **glycoprotein** present in egg white, which has a strong affinity to **biotin**. Can lead to **biotin** deficiency if given in large quantities. Used as a biological reagent in the same way as **streptavidin**.

avidity A measure of the binding strength of an **antibody** to its antigen.

avirulence gene (Abbreviation: *avr* gene). Many plants contain *R* genes, which confer simply-inherited **resistance** to a specific pathogen race. The plants are able to recognize the presence of the pathogen by an interaction between their ***R* gene** and the matching pathogen's avirulence gene. Successful recognition triggers a cascade of further genes, often leading to a **hypersensitive response**.

***avr* gene** Abbreviation for **avirulence gene**.

axenic culture Free of external contaminants and internal symbionts; generally not possible with surface sterilization alone, sometimes used incorrectly to indicate **aseptic culture**.

axillary bud A bud found at the axil of a leaf. *Synonym*: **lateral bud**.

axillary bud proliferation Propagation of plant tissue *in vitro* to promote axillary growth, to generate large numbers of **plantlets** in culture.

B

B cell An important class of **lymphocytes** that mature in bone marrow (in mammals) and the *Bursa of Fabricius* (in birds) and produce **antibodies**. Largely responsible for the antibody-mediated or humoral immune response, giving rise to the antibody-producing **plasma cells** and some other cells of the immune system. *Synonym*: **B lymphocyte**.

B chromosome A supernumerary chromosome present in some individuals (both plant and animal). They are smaller than the normal chromosomes, behave abnormally in both **mitosis** and **meiosis**, can vary in number between somatic cells and are not thought to have any significant gene content.

B lymphocyte *See*: **B cell**.

BABS Abbreviation for **biosynthetic antibody binding sites**.

BAC Abbreviation for **bacterial artificial chromosome**.

bacillus A rod-shaped **bacterium**.

Bacillus thuringiensis (Abbreviation: Bt). A bacterium that produces a toxin against certain insects, particularly *Coloeoptera* and *Lepidoptera*; a major means of **insecticide** for organic farming. Some of the toxin genes are important for **transgenic** approaches to crop protection.

back mutation A second mutation at the same site in a **gene** as the original mutation. The second mutation restores the wild-type **protein** sequence.

backcross Crossing an individual with one of its parents or with the genetically equivalent organism. The **offspring** of such a cross are referred to as the backcross generation or backcross progeny.

bacterial artificial chromosome A **plasmid** vector that can be used to clone large inserts of **DNA** (up to 500 kb). *See*: **yeast artificial chromosome.**

bacterial toxin A toxin produced by a bacterium, such as **Bt toxin** of ***Bacillus thuringiensis***.

bacteriocide A chemical or drug that kills bacterial cells.

bacteriocin A **protein** produced by bacteria of one **strain** and active against those of a closely related strain.

bacteriophage (Abbreviation: phage). A **virus** that infects bacteria. Altered forms are used as cloning **vectors**. *See*: **lambda phage, M13**.

bacteriostat A substance that inhibits or slows down growth and reproduction of bacteria.

bacterium (pl.: bacteria) unicellular prokaryotic organisms, without a distinct nucleus. Major distinctive groups are defined by **Gram staining**. Also classified on the basis of oxygen requirement (aerobic vs anaerobic) and shape (spherical = coccus; rodlike = bacillus; spiral = spirillum; comma-shaped = vibrio; cork-screw-shaped = spirochaete; filamentous).

baculovirus A class of insect virus used to make **DNA cloning vectors** for gene expression in eukaryotic cells. Production of a target **protein** can be up to 50% of the cells' **protein** content, and several **proteins** can be made simultaneously, so that multi-sub-unit enzymes can be made by this system.

baculovirus expression vector (Abbreviation: BEV). A method for the *in vitro* production of complex recombinant eukaryotic **proteins**. A genetically engineered baculovirus (a virus that infects certain types of insects) is introduced into appropriate cultured insect cells, which then express the recombinant **protein**.

balanced lethal system A system for maintaining a recessive lethal **allele** at each of two loci on the same pair of chromosomes. In a closed population with no crossing-over between the loci, only the double heterozygotes for the lethal mutations survive.

balanced polymorphism Two or more **phenotypes** maintained in the same breeding population.

bank *See*: **gene bank**.

bar gene *See*: ***pat*** **gene**.

barnase A bacterial ribonuclease, which, when transformed into plants and expressed in the anthers, generates a male sterile phenotype. Thus, it is a technology applicable to F_1 hybrid seed production, which relies on the ability to genetically sterilize genotypes to ensure that all seed borne on the plant are the result of outcrossing. The sterility phenotype is suppressed by the **barstar protein**, which can therefore be used to reverse the sterility where this is necessary.

Barr body A condensed mass of **chromatin** found in the nuclei of female mammals. It is a late-replicating, inactive X-chromosome. *See:* **dosage compensation, sex linkage**

barstar protein A polypeptide inhibitor of **barnase**.

basal 1. Located at the base of a plant or a plant organ. 2. A fundamental formulation of a tissue culture medium containing nutrients but no growth promoting agents.

base analogue A non-natural **purine** or **pyrimidine** base that differs slightly in structure from the normal bases, but can be incorporated into **nucleic acids**. They are often **mutagenic**.

base One of the components of **nucleosides**, **nucleotides** and **nucleic acids**. Four different bases are found in naturally occurring **DNA** - the **purines** A (**adenine**) and G (**gua-**

nine); and the **pyrimidines** C (**cytosine**) and T (**thymine**, the common name for 5-methyluracil). In **RNA**, T is replaced by U (uracil). *See*: **base pair**.

base pair (Abbreviation: bp). The two separate strands of a nucleic acid **double helix** are held together by specific hydrogen bonding between a **purine** and a **pyrimidine**, one from each strand. The **base** A pairs with T in **DNA** (with U in **RNA**); while G pairs with C in both **DNA** and **RNA**. The length of a nucleic acid molecule is often given in terms of the number of base pairs it contains.

base substitution Replacement of one base by another in a **DNA** molecule. *See*: **transition**; **transversion.**

basic fibroblast growth factor (Abbreviation: BFGF). *See*: **fibroblasts**.

basipetal Developing, in sequence, from the **apex** towards the base. *See*: **acropetal**.

basophil A type of **leukocyte** produced by **stem cells** in the red bone marrow.

batch culture A suspension culture in which cells grow in a finite volume of liquid nutrient medium and follow a sigmoid pattern of growth. All cells are harvested at the same time. *See*: **continuous culture**. *Synonym*: **batch fermentation**.

batch fermentation *See*: **batch culture**.

bench-scale process A small- or laboratory-scale process; commonly used in connection with fermentation.

beta-DNA The form of **DNA** generally found in nature. A right-handed **helix**.

beta-galactosidase A bacterial enzyme that catalyses the cleavage of **lactose** into glucose and galactose, commonly used as a **marker** in **DNA** cloning.

beta-glucuronidase (Abbreviation: GUS). An enzyme produced by certain bacteria, which catalyses the cleavage of a whole range of beta-glucuronides. Because this activity is largely absent in plants, the encoding bacterial gene has been widely used as a **reporter gene** in plant **transgenesis**.

beta-lactamase An enzyme that detoxifies penicillin group **antibiotics**, such as ampicillin. The â-lactamase gene is commonly used as a marker for successful **transformation**, where only transformed cells are able to tolerate the presence of ampicillin. *See*: **selectable marker**.

beta-sitosterol *See*: **phytosterol**.

BEV Abbreviation for **baculovirus expression vector**.

BFGF Abbreviation for **basic fibroblast growth factor**.

biennial A plant which completes its life cycle within two years and then dies.

bifunctional vector *See*: **shuttle vector**.

binary vector system A two **plasmid** system in *Agrobacterium tumefaciens* designed to transfer **T-DNA** into plant cells, while avoiding the formation of **crown gall** tumours. One plasmid contains the virulence gene (responsible for transfer of the T-DNA), and the other T-DNA borders, the selectable marker and the DNA to be transferred.

binding The ability of molecules to bind each other non-covalently because of the exact shape and chemical nature of parts of their surfaces. A common biological phenomenon, as e.g. an **enzyme** to its **substrate**; an **antibody** to its **antigen**; a **DNA** strand to its complementary strand. *See*: **ligand**.

bio- A prefix used in scientific words to associate the concept of "living organisms." Usually written with a hyphen before vowels, for emphasis or in neologisms.

bio-accumulation A problem that can arise when a stable chemical such as a heavy metal or DDT is introduced into a natural environment. Where there are no agents presentable to biodegrade it, its concentration can increase as it passes up the food chain and higher organisms may suffer toxic effects. This phenomenon may be employed beneficially for the removal of toxic metals from wastewater, and for **bioremediation**. *See*: **biosorbents**.

bio-assay 1. The assessment of a substance's activity on living cells or on organisms. Animals have been used extensively in **drug** research in bio-assays in the pharmaceutical and cosmetics industries. Current trends are to develop bio-assays using bacteria or animal or plant cells, as these are easier to handle than whole animals or plants, are cheaper to make and keep, and avoid the ethical problems associated with testing of animals. 2. An indirect method to detect sub-measurable amounts of a specific substance by observing a sample's influence on the growth of live material.

bio-augmentation Increasing the activity of bacteria that decompose pollutants; a technique used in **bioremediation**.

bioavailability The proportion of a nutrient or administered **drug** etc. that can be taken up by an organism in a biologically effective form. For example, some soils high in phosphorus have a low level of P availability because the pH of the soil renders much of the P insoluble.

biocatalysis The use of enzymes to improve the efficiency of chemical reactions.

biochip *See*: **DNA chip**.

biocontrol Pest control by biological means. Any process using deliberately introduced living organisms to restrain the growth and development of other organisms, such as the introduction of predatory insects to control an insect pest. *Synonym*: **biological control**.

bioconversion Conversion of one chemical into another by living organisms, as opposed to their conversion by isolated enzymes or fixed cells, or by chemical processes. Particularly useful for introducing chemical changes at specific points in large and complex molecules.

biodegradable Capable of being biodegraded.

biodegrade The breakdown by microorganisms of a compound to simpler chemicals. Materials that are easily biodegraded are colloquially termed **biodegradable**.

biodesulphurization The removal of organic and inorganic sulphur from coal by bacterial and soil microorganisms. Certain bacteria can oxidize insoluble sulphur compounds into soluble sulphates, which can be washed away with the bacteria. *See*: **bioleaching**.

biodiversity The variability among living organisms from all sources, including, *inter alia*, terrestrial, marine and other ecosystems and the ecological complexes of which they are part; this includes diversity within **species**, between species and of ecosystems. *Synonyms*: **biological diversity, ecological diversity**.

bio-energetics The study of the flow and the transformation of energy that occur in living organisms.

bio-engineering The use of artificial tissues, organs and organ components to replace parts of the body that are damaged, lost or malfunctioning.

bio-enrichment Adding nutrients or oxygen to increase microbial breakdown of pollutants.

bio-ethics The branch of ethics that deals with the life sciences and their potential impact on society.

biofilms A layer of **microorganisms** growing on a surface, in a bed of polymeric material which they themselves have made. Biofilms tend to form wherever a surface on which bacteria can grow is exposed to some suitable medium and a supply of bacteria.

biofuel A gaseous, liquid or solid fuel derived from a biological source, e.g. ethanol, rapeseed oil or fish liver oil.

biogas A mixture of methane and carbon dioxide resulting from the **anaerobic** decomposition of waste such as domestic, industrial and agricultural sewage.

bio-informatics The use and organization of information of biological interest. In particular, concerned with organizing bio-molecular databases (particularly **DNA** sequences), utilizing computers for analysing this information, and integrating information from disparate biological sources. *See*: ***in silico***.

bioleaching The recovery of metals from their ores, using the action of microorganisms, rather than chemical or physical treatment. For example, *Thiobacillus ferroxidans* has been used to extract gold from refractory ores. *See*: **biorecovery**.

biolistics A technique to generate **transgenic**

cells, in which **DNA**-coated small metal particles (tungsten or gold) are propelled by various means fast enough to puncture target cells. Provided that the cell is not irretrievably damaged, the DNA is frequently taken up by the cell. The technique has been successfully used to transform animal, plant and fungal cells, and even mitochondria inside cells. *Synonym*: **microprojectile bombardment**.

biological ageing *See*: **senescence**.

biological containment Restricting the movement of organisms from the laboratory. Can take two forms: making the organism unable to survive in the outside environment, or making the outside environment inhospitable to the organism. For microorganisms, the favoured approach is to engineer organisms to require a supply of a specific nutrient that is usually available only in the laboratory. For higher organisms (plants and animals), it is more possible to ensure that the outside environment is unsuited to growth, spread and reproduction.

biological control *See*: **biocontrol**.

biological diversity *See*: **biodiversity**.

biological oxygen demand (Abbreviation: BOD). The dissolved oxygen required for the respiration of a population of **aerobic** organisms present in water. Expressed in terms of the oxygen consumed in water at a temperature of 20°C per unit time. The BOD is used as an indication of the degree to which the sample of water is polluted, particularly by inorganic nutrients for plants.

biologics Agents, such as **vaccines**, that give immunity to diseases or harmful biotic stresses.

bioluminescence The enzyme-catalyzed production of light by a number of diverse organisms (e.g. fireflies and many deep ocean marine organisms). Utilized as a **reporter gene** in plant **transgenesis**, and for the detection of food-borne pathogenic bacteria.

biomagnification *See*: **bio-accumulation**.

biomass 1. The cell mass produced by a population of living organisms. 2. The organic matter that can be used either as a source of energy or for its chemical components. 3. All the organic matter that derives from the **photosynthetic** conversion of solar energy.

biomass concentration The amount of biological material in a specific volume.

biome A major ecological community or complex of communities, extending over a large geographical area and characterized by a dominant type of vegetation.

biometry The application of statistical methods to the analysis of continuous variation in biological systems. *Synonym*: biometrics.

biomimetic materials Employed to describe synthetic analogues of natural materials with advantageous properties. For instance, some synthetic molecules act chemically like natural **proteins**, but are not as easily degraded by the digestive system. Other systems such as reverse micelles and/or **liposomes** exhibit certain properties that mimic certain aspects of living systems.

biopesticide A compound that kills organisms by virtue of specific biological effects rather than as a broader chemical poison. Differ from **biocontrol** agents in being passive agents, whereas biocontrol agents actively seek the pest. The rationale behind replacing conventional **pesticides** with biopesticides is that the latter are more likely to be selective and **biodegradable.**

biopharming The use of genetically transformed crop plants and livestock animals to produce valuable compounds, especially pharmaceuticals. *Synonym*: molecular pharming.

biopiracy The patenting of genetic stocks, and the subsequent privatization of genetic resources collections. The term implies a lack of consent on the part of the originator.

biopolymer Any large **polymer** (**protein**, nucleic acid, polysaccharide) produced by a living organism. Includes some materials (such as **polyhydroxybutyrate**) suitable for use as plastics. *Synonym*: biological polymer.

bioprocess Any process that uses complete living cells or their components (e.g. enzymes,

chloroplasts) to effect desired physical or chemical changes.

bioreactor A tank in which cells, cell extracts or enzymes carry out a biological reaction. Often refers to a **fermentation** vessel for cells or **microorganisms**.

biorecovery The use of microorganisms for the recovery of valuable materials (metals or particular organic compounds) from complex mixtures. *See*: **biodesulphurization, bioleaching.**

bioremediation A process that uses living organisms to remove contaminants, pollutants or unwanted substances from soil or water. *See*: **remediation, bio-accumulation, bio-augmentation.**

biosafety protocol An internationally agreed protocol set up to protect biological diversity from the potential risks posed by the release of genetically modified organisms. It establishes a procedure for ensuring that countries are provided with the information necessary to make informed decisions before agreeing to the import of such organisms into their territory. *Synonym*: **Cartagena protocol**. *See*: **Convention on biological diversity.**

biosafety Referring to the avoidance of risk to human health and safety, and to the conservation of the environment, as a result of the use for research and commerce of infectious or genetically modified organisms.

biosensor A device that uses an immobilized biologically-related agent (such as an **enzyme**, antibiotic, **organelle** or whole cell) to detect or measure a chemical compound. Reactions between the immobilized agent and the molecule being analysed are converted into an electric signal.

biosilk A **biomimetic** fibre produced by the expression of the relevant orb-weaving spider genes in **yeast** or bacteria, followed by the spinning of the expressed **protein** into a fibre.

biosorbents Microorganisms which, either by themselves or in conjunction with a **substrate** are able to extract and/or concentrate a desired molecule by means of its selective retention. *See*: **bio-accumulation.**

biosphere The part of the earth and its atmosphere that is inhabited by living organisms.

biosynthesis Synthesis of compounds by living cells, which is the essential feature of **anabolism.**

biosynthetic antibody binding sites (Abbreviation: BABS). *See*: **dAb.**

biotechnology 1. "Any technological application that uses biological systems, living organisms, or derivatives thereof, to make or modify products or processes for specific use" (Convention on Biological Diversity). 2. "Interpreted in a narrow sense, ... a range of different molecular technologies such as gene manipulation and gene transfer, **DNA** typing and cloning of plants and animals" (FAO's statement on biotechnology)

biotic factor Other living organisms that are a component of an organism's environment, and form the biotic environment, affecting the organism in many ways.

biotic stress **Stress** resulting from attack by pathogenic organisms.

biotin A vitamin of the B complex, it acts as a **co-enzyme** for various enzymes that catalyse the incorporation of carbon dioxide into various compounds, and is essential for the metabolism of fats. Adequate amounts are normally produced by the intestinal bacteria in animals. Significant as a molecular biology reagent due to its high affinity with **avidin** and **streptavidin**. *Synonym*: **vitamin** H.

biotin labelling The attachment of biotin to another molecule, especially **DNA.**

biotinylated-DNA A **DNA** molecule labelled with **biotin** by incorporation of a biotinylated **nucleotide** (usually **uracil**) into a **DNA** molecule. The detection of the labelled **DNA** is achieved by complexing it with **streptavidin** to which it attached a colour-generating agent such as horseradish peroxidase that gives a fluorescent green colour upon reaction with various organic reagents.

biotope A small habitat in a large community.

biotoxin A naturally produced compound which shows pronounced biological activity, toxic to some or many organisms.

biotransformation The conversion of one chemical or material into another using a biological **catalyst**: a near synonym is biocatalysis, and hence the catalyst used is called a biocatalyst. Usually the catalyst is an **enzyme**, or a fixed whole, dead microorganism that contains an enzyme or several enzymes.

bivalent Two paired homologous chromosomes (one of maternal origin; the other of paternal origin) at **prophase** to **anaphase** of the first meiotic division. Because **DNA** is replicated in prophase, each duplicated chromosome comprises two chromatids, and thus a bivalent comprises four chromatids.

blast cell A large, rapidly dividing cell that develops from a **B cell** in response to an antigenic stimulus. The blast cell then becomes an **antibody**-producing plasma cell.

blastocyst A mammalian embryo (fertilized ovum) in the early stages of development, approximately up to the time of implantation. It consists of a hollow ball of cells.

blastomere Any one of the cells formed from the first few cleavages in animal embryology. The embryo usually divides into two, then four, then eight blastomeres, and so on.

blastula In animals, an early **embryo** form that follows the morula stage; typically, a single-layered sheet (blastoderm) or ball of cells (**blastocyst**).

bleeding 1. Collection of blood from **immunized** animals. 2. Used to describe the occasional purplish-black colouration of media due to phenolic products given off by (usually fresh) transfers.

blot As a verb, to transfer **DNA, RNA** or **protein** to an immobilizing matrix. As a noun, the immobilizing matrix carrying DNA, RNA or protein. The various types of blot are named according to the probe and/or the probed molecules: **Southern blot** (DNA/DNA), **northern blot** (DNA/mRNA), **western blot** (antibody/**protein**), southwestern blot (DNA/protein). Only "Southern" is written with an initial capital, as it is named after Ed Southern, the inventor of the technique.

blunt end The end of a **double-stranded DNA** molecule in which neither strand extends beyond the other. *Synonym*: **flush end.**

blunt-end cut To cut a **double-stranded DNA** with a **restriction endonuclease** which generates **blunt ends.** *Synonym*: **flush-end cut.**

blunt-end ligation The joining of two blunt-ended **double-stranded DNA** molecules.

BOD Abbreviation for **biological oxygen demand.**

boring platform Sterile bottom half of a **Petri dish** used for preparing **explants** with a cork borer.

bound water Cellular water not released into the **intercellular** space upon freezing and thawing. *Opposite*: **free water.**

bovine growth hormone *See*: **bovine somatotrophin**

bovine somatotrophin (Abbreviation: BST) A natural **protein** in cattle. It has been cloned, using **recombinant DNA** technology, expressed in large amounts and marketed as an agricultural product to improve the growth rate and protein:fat ratios in farm cattle, and to improve milk yield. Its use is banned in some countries. *Synonyms*: **bovine growth hormone.**

bovine somatotropin *See*: **bovine somatotrophin.**

bovine spongiform enecelophalopathy (Abbreviation: BSE) Cattle disease (colloquially called mad cow disease) caused by **proteinaceous infectious particles.**

bp Abbreviation for **base pair.**

bract A modified leaf that subtends flowers or inflorescences and may appear to be a petal.

breed 1.a sub-specific group of domestic livestock with definable and identifiable external characteristics that enable it to be separated by visual appraisal from other similarly defined groups within the same species. 2. a group of domestic livestock for which geographical and/or cultural separation from phenotypically similar groups has led to acceptance of its separate identity.

breed at risk An animal breed that is in danger of becoming extinct because its **population** has fallen below a critical number.

breeding The process of **sexual reproduction** and production of offspring.

breeding value A quantitative genetics term, describing that part of the deviation of an individual **phenotype** from the **population** mean that is due to the additive effects of alleles. Thus, if an individual is mated with a random sample of individuals from a population, its breeding value for a given trait is twice the average deviation of its offspring from the population mean for that trait.

brewer's yeast Strains of the yeast *Saccharomyces cerevisiae* that are used for the production of beer.

bridge A filter paper or other substrate used as a wick and support structure for a plant tissue in culture when a **liquid medium** is used.

broad-host-range plasmid A **plasmid** that can replicate in a number of different bacterial species.

broad-sense heritability The proportion of the total phenotypic variation which results from genetic **variation** or interaction between the **genotype** and the environment.

broodstock The group of males and females from which fish are bred.

browning Discolouration of freshly cut surfaces of plant tissue due to phenolic oxidation. In plant tissue culture, it may indicate a nutritional or pathogenic problem, generally leading to **necrosis**.

BSA Abbreviation for bovine **serum albumin**.

BSE Abbreviation for bovine spongiform encephalopathy. *See* **protein**aceous infectious particle.

BST Abbreviation for **bovine somatotrophin**.

Bt Abbreviation for ***Bacillus thuringiensis***.

bubble column fermenter A **bioreactor** in which the cells or microorganisms are kept suspended in a tall cylinder by rising air, which is introduced at the base of the vessel.

bud A region of meristematic tissue with the potential for developing into leaves, shoots, flowers or combinations of these; generally protected by modified scale leaves.

bud sport A **somatic** mutation arising in a bud, generating a genetically different shoot. Includes changes due to gene **mutation**, chromosomal mutation or polyploidy.

budding 1. A method of asexual **reproduction** in which a new individual is derived from an outgrowth (bud) that becomes detached from the body of the parent. 2. Among fungi, budding is characteristic of the brewers yeast *Saccharomyces cerevisiae*. 3. A form of **graft** in which a single vegetative bud is taken from one plant and inserted into **stem** tissue of another plant so that the two will grow together. The inserted bud develops into a new shoot.

bulked segregant analysis A method to obtain **markers** linked to a target **trait**, in which **DNA** samples, prepared from a number of individuals of each of two contrasting phenotypes, are separately pooled and used to generate contrasting **DNA fingerprints**. DNA fragments unique to one pool become candidates for a marker linked to the gene controlling the trait.

buoyant density The intrinsic density which a molecule, virus or sub-cellular particle has when suspended in an aqueous solution of a salt, such as CsCl, or a sugar, such as sucrose. **DNA** from different species has different characteristic buoyant densities, determined largely by the relative proportion of the **base-pairs** G+C, to A+T.

C Abbreviation for **cytosine**.

CAAT box A conserved **DNA** sequence found within the **promoter** region of the **protein**-encoding genes of many eukaryotic organisms. So-called because of its **consensus** sequence GGCCAATCT, it occurs around 75 bases prior to the **transcription** initiation site; and is one of several sites for the recognition and binding of **transcription factors**. *Synonym*: CAT box.

cabinet *See*: growth cabinet.

callipyge An inherited **trait** in livestock (e.g. sheep) that results in thicker, meatier hindquarters, and hence a higher meat yield per animal.

callus (pl.: calli) A protective tissue, consisting of parenchyma cells, that develops over a cut or damaged plant surface. 2. Mass of **undifferentiated**, thin-walled **parenchyma** cells induced by **hormone** treatment. 3. Actively dividing non-organized masses of undifferentiated and differentiated cells often developing from injury (wounding) or in **tissue cul-**

ture in the presence of growth regulators.

callus culture A technique of plant tissue culture, usually on solidified medium and initiated by inoculation of small **explants**. Used as the basis for organogenic (shoot or root forming) cultures, cell cultures or **proliferation** of embryoids. Callus cultures can be indefinitely maintained through regular subculturing.

cambial zone Region in stems and roots consisting of the cambium and its recent derivatives.

cambium (pl.: cambia) A one or two cells thick layer of plant meristematic tissue, between the **xylem** and **phloem** tissues, which gives rise to secondary tissues, thus resulting in an increase in the diameter of the stem or root. The two most important cambia are the **vascular** (fascicular) cambium and the cork cambium.

cAMP Abbreviation for **cyclic adenosine monophosphate**.

CaMV 35S Abbreviation for **cauliflower mosaic virus 35S** ribosomal **DNA promoter**. *See*: **cauliflower mosaic virus.**

CaMV Abbreviation for **cauliflower mosaic virus**.

candidate gene A gene whose deduced function (on the basis of **DNA** sequence) suggests that it may be involved in the genetic control of an aspect of phenotype.

candidate-gene strategy An experimental approach in which knowledge of the biochemistry and/or physiology of a trait is used to identify **candidate genes**. *Synonym*: **functional gene cloning.**

canola A specific subgroup of oilseed rape cultivars; canola oil is the highly mono-unsaturated fatty acid and low in erucic acid product produced in the **seed** of these cultivars.

CAP Abbreviation for **catabolite activator protein**.

cap site The site on a **DNA template** where transcription begins. It corresponds to the **nucleotide** at the 5' end of the **RNA transcript** which accepts the **G cap**.

cap The structure found on the 5′-end of eukaryotic m**RNA**, and consisting of an inverted, methylated **guanosine** residue. *See* **G cap, cap site**.

capacitation The final stage, inside the female genital tract, in the maturation process of a **spermatozoan**, as it penetrates the ovum.

capillary electrophoresis A form of electrophoresis used widely in current large-scale **DNA sequencing** facilities, where the sample is passed through a long, very-narrow-bore tube containing a re-usable matrix.

CAPS *See*: **cleaved amplified polymorphic sequence**.

capsid The **protein** coat of a **virus**. The capsid often determines the shape of the virus. *Synonym*: **coat protein**.

capsule Carbohydrate coverings that have antigenic specificity, present on some types of bacteria and other microorganisms. The capsule is usually composed of polysaccharides, polypeptides, or polysaccharide-**protein** complexes. These materials are arranged in a compact manner around the cell surface.

carbohydrate *See*: polysaccharide.

carboxypeptidase A class of enzymes which catalyse the cleavage of **peptide** bonds, requiring a free carboxyl group in the substrate. The **peptide bond** adjacent to this group is cleaved and a free **amino acid** is released. Used for deriving the amino acid sequence of peptides.

carcinogen A substance capable of inducing cancer in an organism.

carcinoma A malignant tumour derived from epithelial tissue, which forms the skin and the outer cell layers of internal organs.

carotene A reddish-orange **plastid** pigment involved in **photosynthesis**. A **carotenoid** and precursor of **vitamin** A.

carotenoid A group of chemically similar red to yellow pigments responsible for the characteristic colour of many plant organs or fruits, such as tomatoes, carrots, etc. Oxygen-containing carotenoids are called **xanthophylls**. Carotenoids serve as light-harvesting molecules in photosynthetic assemblies and also play a role in protecting prokaryotes from the deleterious effects of light. *See*: **carotene.**

carpel Female reproductive **organ** of flowering plants, consisting of stigma, **style** and ovary.

carrier A **heterozygous** individual bearing a **recessive** mutant **allele** for a defective condition that is "masked" by the presence of the **dominant** normal allele; the phenotype is normal, but the individual passes the defective (recessive) allele to half of its offspring.

carrier DNA DNA of undefined sequence which is added to the transforming **(plasmid) DNA** used in physical **DNA**-transfer procedures. This additional **DNA** increases the efficiency of transformation in **electroporation** and chemically-mediated **DNA** delivery systems. The mechanism responsible is not known.

carrier molecule 1. A molecule that plays a role in moving electrons through the electron transport chain. They are usually **proteins** bound to non-**protein** groups and able to undergo oxidation and reduction relatively easily, thus allowing electrons to flow. 2. A lipid-soluble molecule that can bind to lipid-insoluble molecules and transport them across membranes. Carrier molecules have specific sites that interact with the molecules they transport. The efficiency of carrier molecules may be modified by changing the interacting sites through genetic engineering.

Cartagena protocol *See*: **biosafety protocol.**

casein A group of milk **proteins.**

casein hydrolysate The mixture of **amino acids** and **peptides** produced by enzymatic or acid hydrolysis of **casein.**

cassette *See*: **construct.**

CAT box *See*: **CAAT box.**

catabolic pathway A pathway by which an organic molecule is degraded in order to release energy for growth and other cellular processes.

catabolism The breakdown of large molecules in living organisms, with the accompanying release of energy.

catabolite activator protein (Abbreviation:

CAP). A **protein** which combines with **cyclic AMP**. The cAMP-CAP complex binds to the promoter regions of ***E. coli*** and stimulates transcription of the relevant **operon**. *Synonyms*: catabolite regulator **protein** (CRP), cyclic AMP receptor protein.

catabolite repression Glucose-mediated reduction in the rates of **transcription** of genes that encode enzymes involved in **catabolic pathways** (e.g. the *lac* operon).

catalase A **metalloenzyme**, present in both plants and animals, that catalyzes the decomposition of hydrogen peroxide to water and oxygen. This activity is important in the detoxification of reactive oxygen generated as part of the response to stress.

catalysis The process of increasing the rate of a chemical reaction by the addition of a substance that is not itself changed by the reaction (the **catalyst**).

catalyst A substance that promotes a chemical reaction by lowering the activation energy of a chemical reaction, without itself undergoing any permanent chemical change.

catalytic antibody An antibody selected for its ability to catalyse a chemical reaction by binding to and stabilizing the transition-state intermediate. *Synonym*: abzyme.

catalytic RNA *See*: **ribozyme.**

catalytic site The part of the surface of an **enzyme** molecule (usually only a small portion of the total) necessary for the catalytic process.

cauliflower mosaic virus (Abbreviation CaMV). A **DNA** virus affecting cauliflower and many other **dicot** species. Its importance is due to the **promoter** of its 35S **ribosomal DNA**, which is constitutively active in most plant tissues, and has therefore been widely used as a promoter for the expression of **transgenes**.

cauliflower mosaic virus 35S promoter (Abbreviation CaMV 35S). A promoter sequence isolated from the ribosomal gene of the **cauliflower mosaic virus**.

caulogenesis Stem **organogenesis**; induction of shoot development from **callus**.

CBD Abbreviation for **Convention on Biological Diversity**.

ccc DNA Abbreviation for **covalently-closed circle DNA**. *See*: **circularization**.

CD molecules Abbreviation for: **cluster of differentiation molecules**. Any group of surface **antigens** associated with a specific sub-population of **T cells**.

cDNA Abbreviation for **complementary DNA**.

cDNA clone A double-stranded **cDNA** molecule propagated in a vector, and used as a probe in **RFLP** analyses, as **template** for the production of **EST** sequences, and for **gene expression** studies.

cDNA cloning A method of cloning the coding sequence of a gene, starting with its **mRNA** transcript.

cDNA library An collection of **cDNA** clones.

CDR Abbreviation for **complementarity-determining regions**.

cell culture The *in vitro* growth of cells isolated from multi-cellular organisms.

cell cycle The sequence of stages that a **cell** passes through between one division and the next. The cell cycle oscillates between **mitosis** (M) and the **interphase**, which is divided into the G_1 phase (involving a high rate of biosynthesis and growth), the S phase (in which the **DNA** content is doubled as a consequence of **chromosome** replication), and the G_2 phase (preparatory **for cell division**).

cell differentiation The transition of cells (by the programmed activation and de-activation of the necessary genes) from an tissue-unspecific type, in which daughter cells are similarly undifferentiated, to a committed type in which the **cell line** specializes to become a recognizable tissue or organ.

cell division Formation of two or more daughter cells from a single parent cell. The **nucleus** divides first, followed by the formation of a cell membrane between the daughter nuclei. Division of somatic cells is termed **mitosis**; **egg** and **sperm** precursors are formed following **meiosis**.

cell fusion Formation *in vitro* of a single **hybrid**

cell from the coalescence of two cells of different species origin. In the hybrid cell, the donor nuclei may remain separate, or may fuse, but during subsequent cell divisions, a single **spindle** is formed so that each daughter cell has a single **nucleus** containing complete or partial sets of chromosomes from each parental line. *Synonym*: **cell hybridization**.

cell generation time The interval between the beginning of consecutive divisions of a cell, equivalent to the time that it takes for a population of single-celled organisms to double its cell number.

cell hybridization *See*: **cell fusion**.

cell line 1. A cell **lineage** that can be maintained *in vitro*. Significant genetic changes can occur during lengthy periods in culture, so that the genotype of long-term cell lines may not be the same as that of the starter cell. 2. A cell lineage that can be recognized *in vivo*.

cell membrane *See*: **plasmalemma**.

cell number The number of cells per unit volume of a culture.

cell plate The precursor of the cell wall, formed at the beginning of cell division. The cell plate develops in the region of the equatorial **plate** and arises from membranes in the **cytoplasm**.

cell sap Water and dissolved substances, sugar, amino acids, waste substances, etc., in the plant cell **vacuole**.

cell selection The process of selecting cells exhibiting specific traits within a group of genetically different cells. Selected cells are often sub-cultured onto fresh medium for continued selection and exposed to an increased level of the selection agent to eliminate false positives.

cell sorter *See*: **fluorescence-activated cell sorting, flow cytometry**.

cell strain An *in vitro* culture initiated by asexual reproduction from a single cell. Such cell lines should represent a **population** of genetically homogenous cells. Strains are defined by specific properties or markers used for their selection. *Synonym*: **single-cell line**.

cell suspension Cells in culture in moving or

shaking liquid medium, often used to describe **suspension cultures** of single cells and cell aggregates.

cell The fundamental level of structural organization in complex organisms. Eukaryotic cells contain a **nucleus** (with **chromosomes**) and **cytoplasm** with the **protein** synthesis machinery, bounded by a membrane. Prokaryotic cells have no nucleus.

cell wall A rigid external structure which surrounds plant cells. It is formed outside the **plasmalemma** and consists primarily of **cellulose**.

cell-free protein synthesis *See*: ***in vitro* translation**.

cell-free transcription *See*: ***in vitro* transcription**.

cell-free translation *See*: ***in vitro* translation**.

cell-mediated (cellular) immune response *See*: **T-cell-mediated (cellular) immune response**.

cellular oncogene *See*: **proto-oncogene**.

cellulase Enzyme catalysing the breakdown of **cellulose**.

cellulose A complex **polysaccharide** composed of long linear chains of glucose **residues**. It comprises 40% to 55% by weight of the plant cell wall.

cellulose nitrate *See*: **nitrocellulose**.

cellulosome A multi-**protein** aggregate present in some microorganisms which degrade **cellulose**. It contains multiple copies of the enzymes necessary for this process, and is often found on the outer surface of the **microorganism** cell.

centiMorgan (Abbreviation: cM). Unit of **map distance**. For small **recombination fractions**, cM and % recombination frequency are equivalent.

central dogma The basic concept that, in nature, genetic information generally flows from **DNA** to **RNA** to **protein**. However, information contained in the RNA molecules of **retroviruses** can also flow back to DNA.

central mother cell A subsurface cell located in a plant apical **meristem** and characterized by a large **vacuole**.

centre of origin The geographic locations where

a particular domesticated plant species originated. These areas are the likeliest source of natural genetic variation, and represent ideal targets for ***in situ* conservation**.

centrifugation Separating molecules by size or density using centrifugal forces generated by a spinning rotor. G-forces of several hundred thousand times gravity are generated in ultracentrifugation. *See*: **density gradient centrifugation**.

centrifuge A mechanical device which delivers the centrifugal forces necessary for **centrifugation**.

centriole An **organelle** in many animal cells that appears to be involved in the formation of the **spindle** during **mitosis**. During cell division, the two centrioles move to opposite sides of the **nucleus** to form the ends of the spindle.

centromere The eukaryotic **chromosome** structure, which appears as a constriction in **karyotype** analysis, to which the **spindle** fibres attach during mitotic and meiotic division. Composed of highly **repetitive DNA**.

centrosome A specialized region of a living cell, situated next to the nucleus, where **microtubules** are assembled and broken down during cell division. The centrosome of most animal cells contains a pair of centrioles.

cephem-type antibiotic An antibiotic that shares the basic chemical structure of cephalosporin.

chain terminator 1. *See*: **stop codon**. 2. In the Sanger method of **DNA** sequencing, refers to the labelled di-deoxynucleotide triphosphates which are added to disrupt **DNA polymerase** extension.

Chakrabarty decision A landmark legal case in the U.S.A., in which it was held that the inventor of a new **microorganism** whose invention otherwise met the legal requirements for obtaining a **patent**, could not be denied a patent solely because the invention was alive. This has set the precedent for the patenting of life forms.

chaperone A family of **proteins** that ensure the correct assembly and **conformation** of other polypeptides *in vivo* as they emerge from the **ribosome**, but are not themselves components of the functional assembled structures. The prokaryotic equivalents are known as chaperonins. *See*: **heat shock protein.**

chaperonin *See*: **chaperone.**

character *See*: **trait.**

characterization Description of the essential properties of an organism or system.

charcoal The black porous residue of partly burnt wood, bones, etc; a form of carbon. *See:* **activated charcoal**

chelate A cation bound to an organic molecule through electron pair donation from nitrogen and/or oxygen atoms in its structure. **Ethylenediamine tetraacetic acid** is a typical and frequently employed chelating agent. Soluble chelates can supply plants with **micronutrients** which would otherwise be unavailable because of precipitation.

chemical mutagen A chemical capable of inducing **mutations** in living organisms.

chemically-defined medium When all of the chemical components of a **culture medium** are fully known and defined.

chemiluminescence The emission of light during the course of a chemical reaction.

chemostat A continuous and open **culture** in which **growth rate** and cell density are maintained constant by a fixed rate of input of a growth-limiting nutrient.

chemotaxis The movement of a cell, or the whole or part of an organism, towards or away from an increasing concentration of a particular substance.

chemotherapy The treatment of disease, especially infections or cancer, by means of chemicals.

chiasma (pl.: chiasmata) A visible point of junction between two non-sister **chromatids** of **homologous** chromosomes during the first meiotic prophase. *Synonym*: **cross-over**.

chimera (or chimaera) 1. An organism whose cells are not all genotypically identical. This can occur as a result of: somatic mutation; grafting (*see*: **graft chimera**); or because the individual is derived from two or more embryos or zygotes. 2. A **recombinant DNA** molecule that contains sequences from different organisms.

chimeraplasty A method designed to create defined alterations in **DNA** sequence at a target locus, with potential both for **gene therapy** and for investigating gene function. A synthetic nucleic acid that contains DNA interspersed with small amounts of **RNA** is introduced into the target cell, where it pairs with its target gene sequence and then triggers the cell's **DNA repair** machinery, resulting in the replacement of the native sequence by the synthetic one.

chimeric DNA *See*: **chimera** (2).

chimeric gene An engineered gene, where a coding sequence is fused to **promoter** and/or other sequences derived from a different gene. Most genes used in **transformation** are chimeric. *Synonym*: **fusion gene**.

chimeric protein *See*: **fusion protein**.

chimeric selectable marker gene A gene that is constructed from parts of two or more different genes and allows the host cell to survive under conditions where it would otherwise die.

chip *See*: **micro-array**.

chitin A nitrogenous **polysaccharide** that gives structural strength to the exoskeleton of insects and the cell walls of fungi.

chitinase An **enzyme** which breaks down **chitin**.

chloramphenicol An **antibiotic** that interferes with **protein** synthesis.

chlorenchyma Plant tissue (leaf **mesophyll** and other **parenchyma** cells) containing **chloroplasts**.

chlorophyll One of the two pigments responsible for the green colour of most plants. It is an essential component of the machinery to absorb light energy for photosynthesis. *See*: **chloroplast**.

chloroplast DNA The **DNA** present in the chloroplast. Although, the chloroplast has a small genome, the large number of chloroplasts per cell ensures that chloroplast **DNA** is a significant proportion of the total **DNA** in a plant cell.

chloroplast Specialized **plastid** that contains **chlorophyll**. Lens-shaped and bounded by a double membrane, chloroplasts contain membranous structures (thylakoids) piled up into stacks, surrounded by a gel-like matrix (stroma). They are the site of solar energy transfer and some important reactions involved in starch or sugar synthesis. Chloroplasts have their own **DNA**; these genes are inherited only through the female parent, and are independent of nuclear genes.

chloroplast transit peptide (Abbreviation: CTP). A transit peptide that, when fused to a **protein**, acts to transport that protein into plant chloroplasts. Once inside the chloroplast, the transit peptide is cleaved off the protein. Used to target **transgene** expression to the chloroplast, where this is appropriate.

chlorosis The appearance of yellow colour in plants, due to the failure of development or the breakdown of **chlorophyll**. This is generally a symptom of either nutritional disturbance or of **pathogen** infection.

chromatid Each of the two strands of **chromatin** comprising a duplicated chromosome. The term is applied only while the two chromatids are joined at the **centromere**. As soon as the centromere divides, setting the two chromatids adrift (during **anaphase** of mitosis; and during anaphase II of meiosis), they are called chromosomes.

chromatin fibre The standard structural **conformation** of **chromatin** in strands of 30 nm average diameter.

chromatin Substance of which eukaryotic chromosomes are composed. It consists of a complex of **DNA**, **histone** and **non-histone chromosomal proteins** (mainly histones), and a small amount of **RNA**.

chromatography A method for separating the components of mixtures of molecules by partitioning them between two phases, one stationary and the other mobile. Appropriate selection of partitioning mechanism can produce separation of very closely-related molecules.

chromocentre Body produced by fusion of the heterochromatic regions of the chromosomes in the polytene tissues (e.g. the salivary glands) of certain *Diptera*.

chromogenic substrate A compound or substance that contains a colour-forming group.

chromomeres Small dense bodies identified by their characteristic size and linear arrangement along a chromosome.

chromonema (pl.: chromonemata) An optically single thread forming an axial structure within each **chromosome**.

chromoplast Plastid containing pigments other than **chlorophyll**. *See*: **chloroplast**.

chromosomal aberration An abnormal change in chromosome structure or number, including deficiency, duplication, inversion, translocation, aneuploidy, polyploidy, or any other change from the normal pattern. Although it can be a mechanism for enhancing genetic diversity, most alterations are fatal or debilitating, especially in animals. *See:* **chromosome mutation**.

chromosomal integration site A chromosomal location where **foreign DNA** can be integrated, often without impairing any essential function in the host organism.

chromosomal polymorphism The occurrence of one to several chromosomes in two or more alternative structural forms within a population; the structurally changed chromosomes are the result of chromosome mutations (i.e. any structural change involving the gain, loss or re-location of chromosome segments).

chromosome aberration *See:* **chromosomal aberration**.

chromosome banding Differential staining of chromosomes in such a way that light and dark areas occur along the length of the chromosomes in repeatable patterns. Identical banding pattern implies chromosome homology.

chromosome In eukaryotic cells, chromosomes are the nuclear bodies containing most of the genes largely responsible for the differentiation and activity of the cell. Chromosomes are most easily studied in their contracted state, which occurs around the **metaphase** of **mitosis** or **meiosis**; they contain most of the cell's **DNA** in the form of **chromatin**. Each eukaryotic species has a characteristic number of chromosomes. Bacterial and viral cells contain only one chromosome, which consists of a single or double strand of DNA or, in some viruses, **RNA**, without **histones**.

chromosome jumping A technique that allows two segments of **duplex DNA** that are separated by thousands of base pairs (about 200 kb) to be cloned together. After sub-cloning, each segment can be used as a **probe** to identify cloned **DNA** sequences that, at the chromosome level, are roughly 200 kb apart. *See* **positional cloning.**

chromosome landing An alternative to **chromosome walking** for **positional cloning**. Clones of genomic **DNA** are fragmented so as to include both the target gene and a closely linked marker and are screened to select ('land on') those clones that contain the target gene.

chromosome mutation A change in the gross structure of a chromosome, usually causing severely deleterious effects in the organism, but can be maintained in a population (*See*: **chromosomal polymorphism**). They are often due to meiotic errors. The main types of chromosome mutation are translocation, duplication, and inversion.

chromosome theory of inheritance The theory that chromosomes carry the genetic information and that their behaviour during **meiosis** provides the physical basis for **segregation** and independent assortment.

chromosome walking A strategy for **mapping** or sequencing a chromosome segment and for **positional cloning**. Large restriction **fragments** (or **BAC** clones) are generated and, after probing, a single starting point is identified. New probes are synthesized complementary to sequences of the same fragment (BAC clone) that are adjacent to the starting point, and these are then used to identify different restriction fragments (BAC clones) overlapping the one selected as the starting point. The procedure is used repetitively, working away from the starting point.

chymosin An **enzyme** that clots milk, used in the manufacture of cheese.

ciliate (adj.) *See* **cilium**.

cilium (pl.: cilia) Hairlike locomotor structure on certain cells; a locomotor structure on a ciliate protozoan.

circadian Of physiological activity, etc.: recurring at approximately 24-hour intervals.

circularization The self-ligation of a linear **DNA** fragment having complementary ends, generally generated by digestion with a **restriction endonuclease.** Successful ligation produces a molecule in the form of a covalently-closed circle. **Plastid DNA** and **plasmids** are examples of naturally circularized DNA.

***cis* configuration** *See*: **coupling.**

***cis* heterozygote** A double heterozygote that contains two mutations arranged in a *cis* configuration (e.g. *a+ b+ / a b).*

***cis*-acting protein** A **protein** with the particular property of acting only on the molecule of **DNA** from which it was expressed.

***cis*-acting sequence** A **nucleotide** sequence that only affects the expression of genes located on the same chromosome.

cistron A **DNA** sequence that codes for a specific polypeptide; a gene.

class switching The process during which a **plasma** cell stops producing antibodies of one class and begins producing antibodies of another class.

cleave To break phosphodiester bonds of double-stranded **DNA**, usually with a type II **restriction endonuclease**. *Synonyms*: cut; digest.

cleaved amplified polymorphic sequence A segment of **DNA** that can be amplified by **polymerase chain reaction** (PCR) and which contains a **DNA** sequence **polymorphism**. Following PCR amplification of a locus, the **amplicon** is treated with a **restriction endo-**

nuclease. If the **recognition site** for this enzyme is present in the amplicon, two or more restriction fragments are generated. Thus sequence variation between individuals at the recognition site(s) can be detected by **electrophoresis**. *See also*: **restriction fragment length polymorphism.**

cline Variation in one or more phenotypic characters or allele frequencies across a geographical gradient.

clonal propagation Asexual propagation of many new plants (**ramets**) from an individual (**ortet**); all have the same genotype.

clonal selection The production of a population of **plasma** cells all producing the same **antibody** in response to the interaction between a **B lymphocyte** producing that specific antibody and the **antigen** bound by that antibody. *See*: **primary immune response, secondary immune response**.

clone 1. A group of cells or individuals that are genetically identical as a result of **asexual** reproduction, breeding of completely inbred organisms, or forming genetically identical organisms by nuclear transplantation. 2. Group of plants genetically identical in which all are derived from one selected individual by vegetative propagation. 3. Verb: to clone. To insert a **DNA** segment into a **vector** or host chromosome.

clone bank *See*: **gene bank.**

cloned strain or line A strain or line descended directly from a clone.

cloning *See*: **gene cloning.**

cloning site *See*: **insertion site.**

cloning vector A small, self-replicating **DNA** molecule - usually a **plasmid** or viral **DNA** chromosome - into which foreign **DNA** is inserted in the process of cloning genes or other **DNA** sequences of interest. It can carry inserted **DNA** and be perpetuated in a host cell. *Synonym*: **cloning vehicle.**

cloning vehicle *See*: **cloning vector.**

closed continuous culture A culture system, in which the inflow of fresh medium is balanced by the **outflow** of corresponding volumes of spent medium. Cells are separated mechanically from outflowing medium and added back to the culture.

cluster of differentiation *See:* **CD molecules**.

cM Abbreviation for **centiMorgan**.

CMP Abbreviation for **cytidine monophosphate**. *See*: **cytidylic acid**.

coat protein *See*: **capsid**.

coccus A spherical **bacterium**.

co-cloning The unintentional cloning of **DNA** fragments, along with the desired one, that can occur when the source of **DNA** being cloned is not sufficiently purified.

coconut milk Liquid **endosperm** of the coconut, often used to supply organic nutrients to *in vitro* cultures of plant cells and tissues.

co-culture The joint culture of two or more types of cells, such as a plant cell and a microorganism, or two types of plant cells. Used in various dual-culture systems or in **nurse culture**.

Codex Alimentarius Commission An international regulatory body (part of FAO) responsible for the definition of a set of international food standards. The Commission periodically determines, then publishes a list of food ingredients and maximum allowable levels (the *Codex Alimentarius*) deemed to be safe for human consumption.

coding sequence That portion of a gene which directly specifies the **amino acid** sequence of its product. Non-coding sequences of genes include **introns** and control regions, such as **promoters**, **operators**, and **terminators**.

coding strand The strand of a **DNA** double helix that contains the same base sequence (after substituting U for T) found in the **mRNA** molecule resulting from **transcription** of that segment of **DNA**. Sometimes called the sense strand. The m**RNA** molecule is transcribed from the other strand, known as the **template** or antisense strand. *See*: **antisense DNA**.

coding The specification of a **peptide** sequence, by the code contained in **DNA** molecules. *See*: **genetic code**.

co-dominance Where both **alleles** are expressed

in the heterozygous state, so that the **phenotype** reflects a contribution from both alleles. For example, roan coat colour in cattle results from a mixture of red hairs and white hairs, caused by heterozygosity for the red allele and the white allele.

co-dominant alleles *See*: **co-dominance.**

codon One of the groups of three consecutive nucleotides in **mRNA**, which represent the unit of genetic coding by specifying a particular **amino acid** during the synthesis of polypeptides in a cell. Each codon is recognized by a **tRNA** carrying a specific amino acid, which is incorporated into a **polypeptide** chain during **protein** synthesis. In **DNA**, any informative **triplet** of bases, including both coding and control sequences. *See*: **genetic code, start codon, stop codon**. *Synonym*: triplet. *See*: annex 3.

codon optimization An experimental strategy in which **codons** within a cloned gene - ones not generally used by the host cell **translation** system - are changed by *in vitro* **mutagenesis** to the preferred codons, without changing the amino acids of the synthesized **protein**.

co-enzyme Synonym for **co-factor.**

co-evolution The evolution of complementary adaptations in two species brought about by the **selection** pressure that each exerts on the other. Common in symbiotic associations, in insect-pollinated plants, etc.

co-factor An organic molecule or inorganic ion necessary for the normal catalytic activity of an enzyme. *Synonym*: **co-enzyme.**

co-fermentation The simultaneous growth of two microorganisms in one bioreactor.

cohesive end *See*: **extension.**

coincidence The ratio of the observed to the expected frequency of double **cross-overs**, where the expected frequency is calculated by assuming that the two cross-over events occur independently of one another.

co-integrate A **chimeric DNA** molecule formed by the incorporation at a single site of two different **DNA** molecules.

co-integrate vector system A two **plasmid** system for plant **transgenesis**. One plasmid is engineered to carry a **T-DNA** segment incorporating the gene(s) to be introduced. After introduction into ***Agrobacterium tumefaciens***, the plasmid undergoes homologous recombination with a resident disarmed **Ti plasmid** to form a single plasmid carrying the genetic information for transferring the genetically engineered T-DNA region to plant cells.

colchicine An alkaloid, obtained from the autumn crocus *Colchicum autumnale*, which inhibits **spindle** formation. When applied during **mitosis**, **chromosomes** are unable to separate during **anaphase**. This property is used to achieve a doubling of the chromosome number. A further use is to halt mitosis at metaphase, the stage at which **karyotypes** are best viewed.

coleoptile Protective sheath covering the **shoot apex** of the **embryo** in the grasses.

coleorhiza A protective sheath surrounding the **radicle** in the grasses.

co-linearity 1. A general relationship in which the units in one molecule occur in the same sequence as the units in another molecule which they specify; e.g. the nucleotides in a gene are co-linear with the amino acids in its encoded polypeptide. 2. The phenomenon whereby **gene** order is preserved between distinct species.

collenchyma A tissue of living cells, found particularly in midribs and leaf petioles. Characterized by **cell** walls unevenly thickened with **cellulose** and hemicellulose, but never lignified; it functions as a mechanical support in young, short-lived or non-woody organs.

colony 1. A group of genetically identical cells or individuals derived from a single progenitor. 2. A group of interdependent cells or organisms.

colony hybridization A technique that uses a **nucleic acid probe** to identify a recombinant bacterial **colony** carrying a particular inserted **DNA**.

combinatorial library The many novel combinations (consisting of one heavy and one light **immunoglobulin** chain coding region) that are generated when a heavy-chain **library** is combined by random **pairing** with a light-chain library. These constructs are propagated in a **vector**, and their gene products screened for novel affinity properties.

combining site *See*: **antibody binding site**.

commensalism The interaction of two or more dissimilar organisms where the association is advantageous to one without affecting the other(s).

companion cell A living cell associated with the **sieve cell** of **phloem** tissue in **vascular** plants.

comparative mapping The comparison of map locations of genes and markers between species. In comparisons between closely related species, this will usually uncover a high degree of **conservation** of **synteny** and **co-linearity**. In these cases, the likely location of many genes can be predicted from **model** system data. Comparisons across wider phylogenetic distances reveal increasing loss of synteny.

comparative positional candidate gene Relates to an indirect means of assigning function to a **QTL**. Where a QTL has been linked to a **marker** in one species, and that same marker is linked to a known gene in a **model** system, inferences may be drawn as to the nature of the QTL.

competent 1. Bacterial cells able to take up **foreign DNA** molecules and thereby become genetically transformed. Can be genetically determined, or induced by physical treatment. 2. A competent cell is capable of developing into a fully functional embryo.

complement proteins **Proteins** that bind to **antibody-antigen** complexes and help degrade the complexes by **proteolysis**.

complementarity 1. *See*: **complementary**. 2. The similar correspondence between **DNA** and the **mRNA** transcribed from it.

complementarity-determining regions (Abbreviation: CDR). Regions of the light and heavy subunits of the **immunoglobulin** molecules that interact with the **antigen**. The primary **amino acid** sequences of these regions are highly variable between **antibodies** of the same class. *See*: **antibody binding site**.

complementary DNA (Abbreviation: cDNA). A **DNA** strand synthesized *in vitro* from a mature **RNA** template using **reverse transcriptase**. **DNA polymerase** is then used to create a double-stranded molecule. Differs from genomic DNA by the absence of **introns**. *Synonym*: **copy DNA**.

complementary entity 1. Synonym of **base pair**. 2. One of a pair of segments or strands of **nucleic acid** that will hybridize with one another.

complementary genes Two or more interdependent genes, such that (in the case of dominant complementarity) the **dominant** allele from either gene can only produce an effect on the **phenotype** of an organism if the dominant allele from the other gene is also present; in the case of **recessive** complementarity, only double **homozygous recessive** individuals show the effect.

complementary homopolymeric tailing The process of adding **complementary nucleotide** extensions to **DNA** molecules, (e.g. deoxyguanosine) to the 3' end of one **DNA** molecule and **deoxycytidine** to the 5' end of another **DNA** molecule) to facilitate the **ligation** of the two **DNA** molecules. *Synonyms*: **dA - dT tailing, dG - dC tailing.**

complementary Two **DNA** molecules are complementary to one another when each successive base position from the 5' end in the first molecule is matched by the corresponding residue in the second, starting at the 3' end, according to the normal **base pair** rules (i.e. A for T, C for G). In the appropriate conditions, two complementary **single-stranded DNA** molecules will renature to form a double-stranded molecule. Complementary nucleotides are members of the pairs adenine-thymine, adenine-uracil, and guanine-cytosine that have the ability to hydrogen bond to one another.

complementation *See*: **genetic complementation.**

complementation test A genetic method to test whether or not independent mutations are allelic. In a cross between the two **mutant** individuals, the genotype will be m_1m_2 if the mutations are allelic and $m_1 +/+ m_2$ if non-allelic. The **phenotype** of the former will be mutant, but that of the latter will be **wild type** (normal). *Synonym*: ***trans* test.**

complete digest The treatment of a **DNA** preparation with a **restriction endonuclease** for sufficient time for all of the potential target sites within that **DNA** to have been cleaved. *Opposite*: **partial digest.**

composite transposon A **transposon** formed when two identical or nearly identical transposons insert on either side of a non-transposable segment of **DNA.**

compound chromosome A chromosome formed by the fusion of two separate chromosomes, as in attached-X chromosomes or attached-X-Y chromosomes.

concatemer A **DNA** segment made up of repeated sequences linked head to tail.

concordance Identity of matched pairs or groups for a given trait, such as sibs expressing the same trait.

conditional lethal mutation A **mutation** that is lethal under one set of environmental conditions (the restrictive conditions, commonly associated with high temperature) but is viable under another set of environmental conditions (the permissive conditions).

conditioning 1. The effects on phenotypic characters of external agents during critical developmental stages. 2. The undefined interaction between tissues and **culture medium** resulting in the growth of single cells or small aggregates. Conditioning may be accomplished by immersing cells or callus contained within a porous material (such as **dialysis** tubing) into fresh medium for a period dependent on cell density and a volume related to the amount of fresh medium.

conformation The various three dimensional

shapes that can be adopted by a given molecule. In particular, the different ways in which the **primary sequence** of a biological **polymer** may be folded. This is determined by intra-molecular forces, including hydrogen bonding and, in **proteins**, **disulphide bridges**. In **proteins**, conformation is often critical for biological activity, and the functions of some molecules are carried out by switching between two alternative stable conformations. The native conformation found *in vivo* may be changed to typically less ordered, uncharacterized, and usually biologically-inactive forms by **denaturing**.

conidium (pl.: conidia) An **asexual spore** produced by a specialized hypha in certain fungi.

conjugation 1. Union of **gametes** or **unicellular** organisms during fertilization. 2. The unidirectional transfer of **plasmid DNA** from one bacterium cell to another, involving cell-to-cell contact. The **plasmid** usually encodes the majority of the functions necessary for its own transfer. 3. Attachment of sugar and other polar molecules to less polar compounds, thus making them more water soluble.

conjugative functions Plasmid-based genes and their products that facilitate the transfer of a **plasmid** from one bacterium to another via **conjugation**.

consanguinity Related by descent from a common ancestor.

consensus sequence The part of a gene or **signal sequence** that is shared over a wide range of members of a gene family, both within a given species, or in comparisons between species.

conservation *See:* **gene (resources) conservation, conserved sequence**

conserved sequence An identical or highly similar sequence of **nucleotides** or **amino acids** which occurs as part, or all of a number of different genes or **proteins**, in either the same or different species. This conservation can signify which part of the full sequence is responsible for the functionality.

constant domains Regions of **antibody** chains

that have the same **amino acid** sequence in different members of a particular class of antibody molecules.

constitutive gene A gene that is continually expressed in all cells of an organism.

constitutive promoter An unregulated **promoter** that allows for continual **transcription** of its associated gene.

constitutive synthesis Continual synthesis of a **gene** product by an organism.

constitutive The expression of a **gene** without any requirement for **induction**.

construct An engineered **chimeric DNA** designed to be transferred into a cell or tissue. Typically, the construct comprises the gene or genes of interest, a **marker gene** and appropriate control sequences as a single package. A repeatedly-used construct may be called a cassette.

contained use *See*: **containment**.

containment Measures and protocols applied to limit contact of **genetically modified organisms** or **pathogens** with the external environment. *Synonym*: **contained use**.

contaminant 1. An undesired chemical present in a compound or mixture of compounds. 2. Any **microorganism** accidentally introduced into a **culture** or **culture medium**. The contaminant may compete with the desired cells and consequently inhibit their growth, or totally replace them.

contig A set of overlapping cloned **DNA** fragments that can be assembled to represent a defined region of the **chromosome** or **genome** from which they were obtained. Contig definition is a necessary step for assembling whole genome sequences.

continuous culture A **suspension** culture continuously supplied with nutrients by the inflow of fresh medium. The culture volume is normally constant.

continuous fermentation A process in which cells or microorganisms are maintained in culture in the exponential **growth phase** by the continuous addition of fresh medium that is exactly balanced by the removal of **cell sus-**

pension from the bioreactor.

continuous variation Variation where individuals cannot be classified as belonging to one of a set of discrete classes. Characters showing continuous variation are referred to as quantitative. *See*: **polygene, quantitative trait locus.** *Opposite*: **discontinuous variation**.

controlled environment A closed environment in which parameters, such as light, temperature, relative humidity and sometimes the partial gas pressure (and possibly its composition), are fully controlled.

controlling element In eukaryotes, **transposons** which affect the activity of known genes. This can occur as a result of the **integration** within, or close to a gene, thereby disrupting its activity; or following its **excision** from such a site, thereby restoring activity.

Convention on Biological Diversity (Abbreviation: CBD). The international treaty governing the conservation and use of biological resources around the world, that has also called for the establishment of rules to govern the international movement of non-indigenous living organisms and **genetically modified organisms**.

conversion The development of a **somatic embryo** into a plant.

coordinate repression Correlated regulation of a **structural gene** within an **operon** by a molecule that interacts with the **operator**.

copy DNA *See*: **complementary DNA**.

copy number The number of a particular **plasmid** per bacterium cell, or **gene** per **genome**.

co-repressor An **effector molecule** that forms a complex with a **repressor** and turns off the expression of a gene or set of genes.

corpus A part of the **apical meristem** below the **tunica**. In the corpus, cells divide in all directions, and increase in volume.

correlation A statistical association between variables.

cortex Primary tissue of a stem or root, bounded externally by the **epidermis** and internally in the stem by the **phloem**, and in the root by the pericycle.

cos ends The 12-base, single-strand, complementary **extensions** of **bacteriophage** lambda **DNA**.

cos sites *See* **cos ends.**

co-segregation The joint **inheritance** of two characters, usually the result of genetic **linkage**.

cosmid A synthetic **plasmid** which incorporates the **cos ends**, and one or more **selectable** markers such as an **antibiotic resistance** gene. Cosmids were designed as **vectors** able to incorporate **DNA** fragments up to 40-50 kb in size.

co-suppression A natural **gene silencing** phenomenon, which probably evolved as part of plants' defence against viral attack, but which has become important in the context of plant **transgenesis**. Operates by inhibiting the expression of **transgenes** with homology to native **DNA** through the interaction of native and transgenic **mRNA.**

cot curve A method to estimate the heterogeneity of sequence of a **DNA** preparation, based on the observation that the more homogenous the **DNA**, the more easily (and therefore faster) the annealing of **single-stranded DNA** will occur. The Cot curve plots the extent of annealing from a fully single-stranded preparation over time. The cot (product of initial concentration and time) at which half the DNA has re-natured is the half-cot, a parameter indicating both the degree of heterogeneity in a complex mixture, and of the extent of complementarity in a mixture of two single-stranded DNA molecules.

co-transfection The procedure by which a **baculovirus** and a transfer **vector** are simultaneously introduced into insect cells in culture.

co-transformation A protocol for producing **transgenesis**, in which **host** (plant or animal) cells are transformed simultaneously with two different **plasmids**, one of which carries a **selectable marker**, and the other the gene to be transferred. Relies on the observation that given a sufficiently high concentration of both plasmids, transformed cells will have incorporated both plasmids, possibly at different genomic **loci**. If the transgenes are separable through normal meiotic recombination, transgenic individuals without the selectable marker can be selected in subsequent generations.

cotyledon Leaf-like structures at the first **node** of the seedling stem. In some **dicotyledons**, they represent a food storage organ for the germinating seedling.

coupling The phase state in which either two **dominant** or two **recessive** alleles of two different genes occur on the same chromosome. *Synonym*: ***cis*** **configuration**. *Opposite*: **repulsion; *trans* configuration**.

covalently-closed circular DNA (Abbreviation: ccc DNA). A **DNA** molecule in which the free ends have **ligated** to form a circle. The strands remain linked together even after denaturation. **Plasmids** exist in this form in their *in vivo* state. In its native form, ccc **DNA** will adopt a supercoiled configuration. *See*: **circularization**.

CP4 EPSPS Abbreviation for CP4 5-enolpyruvyl-shikimate-3-phosphate synthase. *See*: **enolpyruvyl-shikimate-3-phosphate synthase**.

cpDNA Abbreviation for **chloroplast DNA**.

cross hybridization The annealing of a single-stranded **DNA** sequence to a single-stranded target **DNA** to which it is only partially complementary. Often, this refers to the use of a **DNA probe** to detect **homologous** sequences in species other than the origin of the **probe**.

cross pollination Application of **pollen** from one plant to another to effect the latter's fertilization.

cross pollination efficiency The ease with which **cross pollination** can be achieved. Generally measured by the number of **hybrid** progeny generated per flower pollinated.

cross The mating of two individuals or populations. *See*: **cross-breeding**.

cross-breeding Mating between members of

different populations (lines, breeds, races or species). *See*: **cross**.

crossing over The process by which **homologous** chromosomes exchange material at **meiosis** through the breakage and reunion of non-sister chromatids. *See:* **recombination, chiasma**.

crossing-over unit *See*: **recombination fraction**.

cross-over *See*: **chiasma**.

crown gall A tumorous growth at the base of certain plants characteristic of infection by ***Agrobacterium tumefaciens***. The **gall** is induced by the **transformation** of the plant cell by portions of the **Ti plasmid**.

crown The base of the **stem** of cereals and forage species from which tillers or branches arise. In woody plants, the root-stem junction. In forestry, the top portion of the tree.

CRP Abbreviation for catabolite regulator protein. *See*: **catabolite activator protein**.

cry proteins A class of crystalline **proteins** produced by strains of ***Bacillus thuringiensis***, and engineered into crop plants to give **resistance** against insect pests. These **proteins** are toxic to certain categories of insects (e.g. corn borers, corn rootworms, mosquitoes, black flies, armyworms, tobacco hornworms, some types of beetles, etc.), but are harmless to mammals and most beneficial insects. *Synonym*: **delta endotoxins**.

cryobiological preservation The preservation of **germplasm** resources in a dormant state by storage at ultra-low temperatures, often in liquid nitrogen. Currently applied to storage of plant seeds and **pollen**, microorganisms, animal **sperm**, and **tissue culture** cell lines. *Synonyms*: **cryopreservation, freeze preservation**.

cryogenic At very low temperature.

cryopreservation *See*: **cryobiological preservation**.

cryoprotectant Compound preventing cell damage during successive freezing and thawing processes. Cryoprotectants are agents with high water solubility and low toxicity. Two types commonly used: permeating (glycerol and **DMSO**) and non-permeating (sugars, dextran, ethylene glycol, **polyvinylpyrrolidone** and hydroxyethyl starch).

cryptic Anything hidden. 1. Structurally **heterozygous** individuals that are not identifiable as they do not show abnormal meiotic chromosome pairing configurations ('cryptic structural hybrids'). 2. A form of polymorphism controlled by recessive genes ('cryptic **polymorphism**'). 3. Any **mutation** which is exposed by a sensitizing mutation and otherwise poorly detected (such mutations probably escape detection because of the plasticity of composition of the corresponding polypeptide). 4. Phenotypically very similar species (cryptic species) which do not hybridize under normal conditions. 5. Cryptic genetic variation refers to the existence of, for example, alleles conferring high performance for a **trait**, in a breed that has low performance for that trait.

CTP Abbreviation for 1. **cytidine** 5'-triphosphate, which is required for **RNA** synthesis since it is a direct precursor molecule; 2. **Chloroplast transit peptide**.

cultigen A cultivated plant **species** with no known wild progenitor.

cultivar (Abbreviation: cv). An internationally accepted term denoting a **variety** of a cultivated plant. Must be distinguishable from other varieties by stated characteristics and must retain their distinguishing characters when reproduced under specific conditions.

culture A **population** of plant or animal cells or microorganisms grown under controlled conditions.

culture alteration A term used to indicate a persistent change in the properties of a culture's behaviour (e.g. altered morphology, **chromosome** constitution, virus susceptibility, nutritional requirements, proliferative capacity, etc.). The term should always be qualified by a precise description of the change which has occurred in the culture.

culture medium Any nutrient system for the cultivation of cells, bacteria or other organisms; usually a complex mixture of organic and inorganic nutrients.

culture room A dedicated room for maintaining **cultures**, often in a **controlled environment**.

curing The elimination of a **plasmid** from its host cell. Many agents which interfere with **DNA** replication, e.g. **ethidium bromide**, can cure plasmids from either bacterial or eukaryotic cells.

cut *See*: **cleave**.

cuticle Layer of cutin or wax, formed on the outer surface of leaves and fruits, thought to have evolved to reduce evaporative water loss.

cutting A detached plant part that, with appropriate treatment, can regenerate into a complete plant.

cybrid A hybrid, originating from the fusion of a cytoplast (the **cytoplasm** without nucleus) with a whole cell derived from a different species.

cyclic adenosine monophosphate (Abbreviations: cyclic AMP, cAMP). A "messenger" molecule that regulates many intracellular reactions by transducing signals from extracellular growth factors to cellular metabolic pathways.

cyclic AMP Abbreviation for **cyclic adenosine monophosphate**.

cyclodextrin Cyclic **oligomer** of glucose.

cycloheximide A molecule that inhibits **protein** synthesis in eukaryotes, but not in prokaryotes. It blocks **peptide bond** formation by binding to the large ribosomal **subunits**. *Synonym*: actidione.

cytidine The (ribo)**nucleoside** resulting from the combination of the **base cytosine** (C) and the sugar **D-ribose**. The corresponding **deoxyribonucleoside** is called deoxycytidine. *See*: **CTP** (1), **dCTP**, **cytidylic acid**.

cytidine triphosphate (cytidine 5'-triphosphate) *See*: **CTP** (1).

cytidylic acid Synonym for **cytidine** monophosphate (abbreviation: CMP), a (ribo)**nucleotide** containing the **nucleoside** cytidine. The corresponding deoxyribonucleotide is called deoxycytidine 5'-monophosphate or deoxycytidylic acid.

cytochrome A class of pigments in plant and animal cells, usually in the mitochondria. They function as electron carriers in respiration.

cytochrome p450 A highly diversified set (more than 1500 known sequences) of heme-containing **proteins**. Frequently called hydroxylases, although P450 **proteins** can perform a wide spectrum of other reactions. In bacteria, they are soluble and approximately 400 amino acids long; eukaryotic P450s are larger - about 500 amino acids. In mammals, they are critical for drug metabolism, haemostasis, cholesterol **biosynthesis** and steroidogenesis; in plants they are involved in plant, **hormone** synthesis, phytoalexin synthesis, flower **petal pigment** biosynthesis and many unknown functions. In fungi, they make ergosterol and they are involved in pathogenesis. Bacterial P450s are key elements in **antibiotic** synthesis.

cytogenetics The biology of chromosomes and their relation to the transmission and **recombination** of genes.

cytokine A generic name for a diverse group of soluble **proteins** and peptides which act as humoral regulators at extremely small concentrations and which, either under normal or pathological conditions, modulate the functional activities of individual cells and tissues. *See*: **lymphokine, monokine**.

cytokinesis Cytoplasmic division and other changes exclusive of nuclear division that are a part of **mitosis** or **meiosis**.

cytokinin Plant growth regulators characterized as substances that induce **cell division** and cell differentiation. In tissue culture, these substances are associated with enhanced callus and shoot development. The compounds are derivatives of **adenine**. *See*: **kinin**.

cytology The study of the structure and function of cells.

cytolysis Cell disintegration.

cytoplasm The living material of the cell, exclusive of the nucleus, consisting of a complex **protein** matrix or gel, and where essential membranes and cellular organelles (mitochondria, plastids, etc.) reside.

cytoplasmic genes Genes located on **DNA** outside the nucleus, i.e. on **plastids.**

cytoplasmic inheritance Hereditary transmission dependent on **cytoplasmic genes.**

cytoplasmic male sterility Genetic defect due to faulty functioning of mitochondria in **pollen** development, preventing the formation of viable pollen. Commonly found or **inducible** in many plant species and exploited for some F_1 **hybrid** seed programmes.

cytoplasmic organelles Discrete sub-cellular structures located in the cytoplasm of cells - mitochondria, **plastids** and **lysosomes.**

cytosine (Abbreviation: C). One the **bases** found in **DNA** and **RNA**. *See*: **cytidine.**

cytosol The fluid portion of the **cytoplasm**, i.e. the cytoplasm minus its organelles.

cytotoxic T cell *See*: **killer T cell.**

cytotoxicity Poisoning of the cell.

cytotype A maternally inherited cellular condition in *Drosophila* that regulates the activity of transposable P elements.

D loop Abbreviation for displacement loop. Formed when a short stretch of **RNA** is paired with one strand of **DNA**. This displaces the original partner DNA; also the displacement of a region of one strand of **duplex DNA** by a single-stranded invader in the reaction catalyzed by **recA**.

dA - dT tailing *See* **complementary homopolymeric tailing**.

dAb (Full term: **single domain antibody**). **Antibodies** with only one (instead of two) **protein** chain derived from only one of the two domains of the normal antibody structure. Exploits the finding that for some antibodies, half of the molecule binds to its target **antigen** almost as well as the whole molecule. The major advantage of dAbs over other antibodies is that they can be cloned and expressed into bacteria, so that large numbers of antibodies can be generated and screened in parallel.

DAF *See*: **DNA amplification fingerprinting**.

Dalton (Abbreviation: Da). A unit of atomic

mass roughly equivalent to the mass of a hydrogen atom. Used as to express molecular weight, which for biological **macromolecules** is usually in the range kilo- (kDa) to megaDaltons (MDa).

DAMD *See*: **directed amplification of minisatellite DNA**.

Darwinian cloning Selection of a **clone** from a large number of essentially random starting points, rather than isolating a natural **gene** or making a carefully designed artificial one. Molecules which are more similar to those needed are selected, mutated to generate new variants, and re-selected. The cycle proceeds until the required molecule is found. The advantage of the system is that the selection is from a vast number of possibilities.

dATP Abbreviation for deoxyadenosine 5'-triphosphate. dATP is required for **DNA** synthesis since it is a direct precursor molecule. *See*: **adenosine**, **adenylic acid**.

dCTP Abbreviation for deoxycytidine 5'-triphosphate. dCTP is required for **DNA** synthesis since it is a direct precursor molecule. *See*: **cytidine**, **cytidylic acid**.

ddNTP Abbreviation for **di-deoxynucleotide**.

death phase The final **growth phase** of cell culture, during which nutrients have been depleted and **cell number** decreases.

deceleration phase The phase of declining growth rate, following the **linear phase** and preceding the **stationary phase** in most batch-suspension cultures. *See*: **growth phase**.

de-differentiation The process, in response to wounding and in tissue cultures, by which plant cells can become unspecialized and start to proliferate by **cell division** to form a mass of **undifferentiated** cells (or **callus**) which, in response to appropriate stimuli, may later differentiate again to form either the same cell type or a different one.

defective virus A virus that, by itself, is unable to reproduce when infecting its **host** cell, but that can grow in the presence of another virus. This other virus provides the necessary molecular machinery that the first virus lacks.

deficiency Lack of adequate supply of nutritional, enzymatic, or environmental requirements, so that development, growth or physiological functions are affected.

defined 1. Fixed conditions of medium, environment and **protocol** for growth. 2. Precisely known and stated elements of **a tissue culture** medium.

degeneracy The specification of one **amino acid** by more than one **codon**. It arises from the inevitable redundancy resulting from the 64 possible codons encoding only 20 amino acids.

degeneration 1. Changes in cells, tissues or organs due to disease. 2. The reduction in size or complete loss of organs during evolution.

dehalogenation The removal of halogen atoms (fluorine, chlorine, bromine, iodine) from molecules, for example during biodegradation.

dehiscence The spontaneous and often violent opening of a fruit, **seed** pod or **anther** to release and disperse the seeds or pollen.

dehydrogenase An **enzyme** that catalyses the removal of hydrogen atoms in biological reactions.

dehydrogenation A chemical reaction in which hydrogen is removed from a compound.

de-ionized water Water from which most salts have been removed - with varying degrees of efficiency - by ion exchange.

deletion A **mutation** involving the removal of one or more **base** pairs in a **DNA** sequence. Large deletions are sometimes microscopically visible in **karyotype** analyses.

deliberate release In a biotechnology context, the intentional release of **genetically modified organisms**.

delta endotoxins *See*: ***cry* proteins**.

deme A group of organisms in the same taxon.

demineralize To remove the mineral content (salts, ions) from a substance, especially water. Removal methods include distillation, electrodialysis and ion exchange. *See*: **de-ionized water**.

denature To disrupt the normal *in vivo* **confor-**

mation of a nucleic acid or (more usually) a **protein** by physical or chemical means, usually accompanied by the loss of activity. *See*: **denatured DNA, denatured protein.**

denatured DNA Double-stranded **DNA** that has been converted to single strands by breaking the hydrogen bonds linking **complementary nucleotide** pairs. Often reversible. Usually achieved by heating.

denatured protein Altering the *in vivo* **conformation** of a **protein** by heat or salt treatment, thereby destroying its biological activity. Unlike denatured **DNA**, denatured **proteins** are seldom able to be renatured.

denaturing gradient gel electrophoresis (Abbreviation: **DGGE**). An **electrophoresis** method for separating similar sized **DNA** fragments on the basis of their sequence, by applying across the gel a gradient of increasingly denaturing conditions (usually by increasing the concentration of a denaturing chemical, such as formamide or urea). As the double-stranded molecules denature into a partially and eventually a fully single-stranded state, their electrophoretic mobility changes.

dendrimer A **polymer** that repeatedly branches until stopped by the physical constraint of having formed a complete, hollow sphere. These structures possess sites on their exterior surface to which **DNA** fragments can be attached, and are thus useful as carriers of DNA for transgenesis.

denitrification A chemical process in which nitrates in the soil are reduced to molecular nitrogen, which is released to the atmosphere.

density gradient centrifugation High-speed **centrifugation** in which molecules are separated on the basis of their different densities using a concentration gradient of caesium chloride or sucrose. The density gradient may either be formed before centrifugation by mixing two solutions of different density (as in sucrose density gradients) or it can be formed by the process of centrifugation itself (as in CsCl and Cs_2SO_4 density gradients).

deoxyadenosine *See*: **adenosine, dATP.**
deoxycytidine *See*: **cytidine, dCTP.**
deoxyguanosine *See*: **guanosine, dGTP.**
deoxyribonuclease *See*: **DNase.**
deoxyribonucleic acid *See*: **DNA.**
deoxyribonucleoside *See*: **nucleoside.**
deoxyribonucleotide *See*: **nucleotide.**
deoxyribose (2-deoxyribose) *See*: **ribose.**
deoxythymidine Strictly correct but rarely used synonym for **thymidine**.
derepression The process of "turning on" the expression of a gene or set of genes whose expression has been repressed (turned off), usually by the displacement of a **repressor** from a **promoter**, since, when attached to the **DNA**, the repressor prevents **transcription.**
derivative 1. Resulting from or derived from. 2. Term used to identify a variant during meristematic **cell** division.
desiccant Any compound used to remove moisture or water.
desoxyribonucleic acid Obsolete spelling of **deoxyribonucleic acid.**
desulphurization *See*: **biodesulphurization.**
detergent Substance which lowers the surface tension of a solution, improving its cleaning properties.
determinate growth Growth determined and limited in time, with a **bud** or flower terminating the growth of the main axis. Once established, it is usually irreversible. *Opposite*: **indeterminate growth.**
determination Process by which undifferentiated cells in an **embryo** become committed to develop into specific **cell** types, such as neurons, fibroblasts or muscle cells.
determined Describing embryonic tissue at a stage when it can develop only as a certain kind of tissue.
development The sum total of events that contribute to the progressive elaboration of an organism. The two major aspects of development are growth and differentiation.
deviation 1. An alteration from the typical form, function or behaviour. **Mutation** or **stress** are the common reasons behind deviation. 2. A

statistical term describing the difference between an actual observation and the **mean** of all observations.

dextrin An intermediate **polysaccharide** compound resulting from the **hydrolysis** of starch to maltose by **amylase** enzymes.

dG - dC tailing *See*: **complementary homopolymeric tailing.**

DGGE *See*: **denaturing gradient gel electrophoresis.**

dGTP Abbreviation for deoxyguanosine 5'-triphosphate. dGTP is required for **DNA** synthesis since it is a direct precursor molecule. *See*: **guanosine, guanylic acid.**

diagnostic procedure A test or **assay** used to determine the presence of a specific substance, organism or **nucleic acid** sequence alteration, etc.

diakinesis A stage of **meiosis** at the end of **prophase** I, in which the contraction of the chromosomes is almost at a maximum, pairing configurations are well defined, the nucleolus normally disappears and the nuclear envelope is disrupted.

dialysis A biochemical technique by which large molecules such as **proteins** in solution are separated from smaller species such as salts. The technique is based on the properties of certain membrane structures, which selectively only allow the passage of the smaller molecules. A frequently used method for the purification of proteins.

diazotroph An organism that can fix atmospheric nitrogen.

dicentric chromosome A chromosome having two active **centromeres.**

dichogamy The condition in which the male and the female reproductive organs of a flower (or certain hermaphroditic animals) mature at different times, thereby making self-fertilization improbable or impossible.

dicot *See*: **dicotyledon.**

dicotyledon (Abbreviation: dicot). A plant with two **cotyledons**. One of the two major classes of flowering plants (along with the **monocotyledons**). Examples include many crop plants (potato, pea, beans), ornamentals (rose, ivy) and timber trees (oak, beech, lime).

di-deoxynucleotide (Abbreviations: ddNTP,didN). A synthetic **deoxynucleotide** that lacks a 3'-hydroxyl group, and is thus unable to form the 3'?5' **phosphodiester bond** necessary for chain elongation. Used as strand terminators in the Sanger **DNA** sequencing reaction and in the treatment of some viral diseases.

didN *See*: **di-deoxynucleotide**.

differential centrifugation A method for separating sub-cellular particles according to their sedimentation coefficients, which are roughly proportional to their size. Cell extracts are subjected to a succession of **centrifuge** runs at progressively faster rotation speeds. Large particles, such as nuclei or mitochondria, will be precipitated at relatively slow speeds; higher G forces will be required to sediment small particles, such as **ribosomes**.

differential display A method to identify **mRNAs** which are present at different levels in different tissues, or in response to specific treatments. The m**RNA**s are converted to **cDNA**, and a defined proportion of these are amplified by the **polymerase chain reaction**, and separated by **electrophoresis**.

differentially permeable Referring to a membrane, through which different substances diffuse at different rates. Some substances may be unable to diffuse through such a membrane, usually because they are too large to fit through the pores of the membrane.

differentiation A process as a result of which unspecialized cells develop structures and functions characteristic of a particular type of **cell**, typically during the process of **development** from one cell to many cells, accompanied by a modification of the new cells for the performance of particular functions. The process is generally irreversible *in vivo* in higher organisms. In tissue culture, the term is used to describe the formation of different cell types.

diffusion The spontaneous movement of molecules from a region of higher concentration to a region of lower concentration.

digest To treat **DNA** molecules with one or more **restriction endonucleases** in order to **cleave** them into smaller fragments.

dihaploid An individual which arises from a doubled **haploid**.

dihybrid An individual that is heterozygous for two pairs of alleles; the **progeny** of a **cross** between homozygous parents differing at two loci.

dimer 1. A molecule formed by the covalent combination of two **monomers**, generally accompanied by elimination of water. 2. The reversible association of two similar (or nearly similar) molecules. The active form of many enzymes is as a dimer between two non-active monomeric subunits.

dimethyl sulphoxide (Abbreviation: DMSO). A highly hygroscopic liquid and powerful solvent with little odour, colour or **toxicity** when pure. It is employed in small quantities to dissolve organic substances in **tissue culture** media preparation and has uses as a **cryoprotectant** and in promoting the passage of chemicals through skin.

dimorphism The existence of two distinctly different types of individuals within a species. An obvious example is sexual dimorphism in mammals.

dinucleotide A nucleotide **dimer**.

dioecious A plant **species** in which male and female flowers form on different plants.

diplochromosome *See*: **endoreduplication**.

diploid The status of having two complete sets of **chromosomes**, most commonly one set of paternal origin and the other of maternal origin. **Somatic** tissues of higher plants and animals are ordinarily diploid in chromosome constitution, in contrast with the **haploid gametes**.

diplonema Stage in **prophase** I of **meiosis** following the **pachytene** stage, but preceding **diakinesis**, in which one pair of sister chromatids begin to separate from the other pair.

diplotene (adj.) *See* **diplonema**.

direct embryogenesis The formation in culture, on the surface of zygotic or **somatic** embryos or on **explant** tissues (leaf section, root tip, etc.), of embryoids without an intervening **callus** phase. *Opposite*: **indirect embryogenesis**.

direct organogenesis Formation of organs directly on the surface of cultured intact **explants**. The process does not involve **callus** formation. *Opposite*: **indirect organogenesis**.

direct repeat Two or more stretches of **DNA** within a single molecule which have the same **nucleotide sequence** in the same orientation. Direct repeats may be either adjacent to one another or far apart on the same molecule.

directed amplification of minisatellite DNA (Abbreviation: DAMD). A **polymerase chain reaction** technique used for obtaining molecular markers in the region of **minisatellites**. To target these regions, one of the **primers** is directed to a **VNTR** core sequence.

directed mutagenesis The generation of changes in the **nucleotide sequence** of a cloned **gene** by one of several procedures. Undertaken to explore the relationship between nucleotide sequence and gene function, and to modify gene products. *Synonym*: ***in vitro* mutagenesis**.

directional cloning The technique by which a vector and a **DNA** insert are both digested with two different **restriction endonucleases** to create non-complementary **sticky ends** at either end of both molecules, so favouring the **insert** to be ligated into the **vector** in a specific orientation, while also preventing the vector from re-circularizing.

disaccharide A **dimer** consisting of two covalently linked **monosaccharides**.

disarm The **deletion** from a **plasmid** or **virus** of genes that are pathogenic.

discontinuous variation Variation where individuals can be classified as belonging to one of a set of discrete, non-overlapping classes. Generated by simple genetic control of a **trait** (one or a small number of genes, each of large effect) and involving minimal non-genetic effect. Characters showing discontinuous variation are referred to as qualitative. *Opposite*: **continuous variation**.

discordant Members of a pair showing different, rather than similar, characteristics.

disease resistance The genetically determined ability to prevent the reproduction of a **pathogen**, thereby remaining healthy. Some resistances operate by pathogen exclusion, some by preventing pathogen spread, and some by tolerating pathogen **toxin**.

disease-free A plant or animal certified through specific tests as being free of specified **pathogens**. Should be interpreted to mean "free from any *known* disease" as "new" diseases may yet be discovered to be present.

disease-indexing Disease-indexed organisms have been assayed for the presence of known diseases according to standard testing procedures.

disinfection Attempted elimination by chemical means of internal microorganisms (particularly pathogens) from a **culture** or sample; rarely attained. *See*: **sterilize** (1).

disinfestation The elimination or inhibition of the activity of surface-adhering microorganisms and removal of insects.

disjunction Separation of **homologous** chromosomes during **anaphase** I of **meiosis**, or of sister chromatids during anaphase of **mitosis** and anaphase II of meiosis.

disomic (adj.) *See* **disomy**.

disomy The presence of a pair of a specific homologous chromosomes. This is the norm for **diploids**.

dispense The transfer of a measured volume of a solution.

disrupter gene Used to enforce the sterility of seed saved from a genetically engineered crop. *See*: **genetic use restriction technology**.

dissecting microscope A microscope with a magnifying power of about 50x, used as an aid in the manipulation of small objects, e.g. **excision** of embryos from young zygotes.

dissection Separation of a tissue by cutting into components, for analysis or observation.

distillation The process of heating a mixture to separate the more volatile from the less volatile parts, and then condensing fractions of the resulting vapour so as to produce a more nearly pure or refined substance.

disulphide bond *See*: **disulphide bridge**.

disulphide bridge A chemical bond between pairs of sulphur atoms that stabilizes the three-dimensional structure of proteins, and hence the **protein**'s normal function. These form particularly readily between cysteine **residues** in the same or different **peptide** molecules. *Synonym*: **disulphide bond**.

ditype In fungi, a **tetrad** that contains two kinds of meiotic products (spores), e.g. 2AB and 2ab.

diurnal An event that occurs repetitively on a daily basis, generally during daylight hours.

dizygotic twins Two-egg twins, i.e. a pair of individuals that shared the same uterus at the same time, but which arose from separate and independent **fertilization** of two ova.

DMSO *See*: **dimethyl sulphoxide**.

DNA Abbreviation for deoxyribonucleic acid, former spelling desoxyribonucleic acid. A long chain polymer of **deoxyribonucleotides**. **DNA** constitutes the genetic material of most known organisms and organelles, and usually is in the form of a **double helix**, although some viral genomes consist of a single strand of **DNA**, and others of a single- or a double-stranded **RNA**. *See*: **base pair, genetic code**.

DNA amplification fingerprinting (Abbreviation: DAF). An **arbitrarily primed polymerase chain reaction** technique for obtaining molecular markers using very short (5-8 bp) **primers**.

DNA amplification Many-fold multiplication of a particular **DNA sequence** either *in vivo* in a **plasmid, phage** or other **vector**; or *in vitro* using, most commonly, the **polymerase chain reaction**.

DNA chip *See*: **micro-array**.

DNA cloning *See*: **gene cloning**.

DNA construct A chimeric **DNA** molecule, carrying all the genetic information necessary for its **transgenic** expression in a host cell.

DNA delivery system A generic term for any procedure that transports **DNA** into a recipient cell.

DNA diagnostics The use of **DNA** polymorphisms to detect the presence of a specific sequence, which could indicate the presence of a contaminant, of a pathogen, or of a particular allele at a target gene. Most commonly utilises the **polymerase chain reaction.**

DNA fingerprint A description of the **genotype** of an individual from the pattern of **DNA** fragments obtained from **DNA fingerprinting**. *Synonym*: **DNA profile**.

DNA fingerprinting The derivation of unique patterns of **DNA** fragments obtained using a number of marker techniques; historically these were **RFLPs,** but latterly they are generally **polymerase chain reaction** based. *Synonym*: **genetic fingerprinting.**

DNA helicase An enzyme that catalyses the unwinding of the complementary strands of a **DNA** double helix. *Synonym*: **gyrase**.

DNA hybridization The annealing of two **single-stranded DNA** molecules, possibly of different origin, to form a partial or complete **double helix**. The degree of hybridization varies with the extent of **complementarity** between the two molecules, and this is exploited to test for the presence of a specific **nucleotide** sequence in a **DNA** sample.

DNA ligase An enzyme that catalyses a reaction to link two separate **DNA** molecules via the formation of a **phosphodiester bond** between the 3'-hydroxyl end of one and the 5'-phosphate of the other. Its natural role lies in **DNA repair** and replication. An essential tool in recombinant **DNA** technology, as it enables the incorporation of foreign DNA into vectors.

DNA micro-array *See:* **micro-array, somatic cell hybrid panel, radiation hybrid cell panel**

DNA polymerase *See*: **polymerase**.

DNA polymorphism The existence of two or more alternative **alleles** at a **DNA**-based marker locus.

DNA primase An enzyme that catalyses the synthesis of the short strands of **RNA** that initiate the synthesis of **DNA** strands.

DNA probe *See*: **probe**.

DNA profile *See*: **DNA fingerprint**.

DNA repair A variety of mechanisms that repair errors (e.g. the incorporation of a non-complementary nucleotide) that occur naturally during **DNA replication**.

DNA replication The process whereby **DNA** copies itself, under the action of and control of **DNA polymerase**.

DNA sequencing Procedures for determining the nucleotide sequence of a **DNA** fragment. Two common methods available: 1. The Maxam Gilbert technique, which uses chemicals to **cleave DNA** into fragments at specific bases; or, most commonly, 2. the Sanger technique (also called the di-deoxy or chain-terminating method) which uses DNA **polymerase** to make new DNA chains, in the presence of **di-deoxynucleotides** (chain terminators) to stop the chain randomly as it grows. In both cases, the **DNA** fragments are separated according to length by **polyacrylamide gel electrophoresis**, enabling the sequence to be read directly from the gel. The procedure has become increasingly automated and large-scale in recent years.

DNA topo-isomerase An enzyme that catalyses the introduction or removal of supercoils in **DNA**. *Synonym*: **topo-isomerase**.

DNA transformation *See*: **transformation**.

DNA vaccine A vaccine generated by the injection of specific **DNA** fragments to stimulate an immune response.

DNAase *See* **DNAse**.

DNAse Abbreviation for deoxyribonuclease. Any enzyme that catalyses the cleavage of DNA **phosphodiester bonds**. DNAse I is a digestive **endonuclease** secreted by the pancreas, that degrades **DNA** into shorter fragments. Many other endonucleases and **exonucleases** are involved in **DNA repair** and replication. *Synonym*: DNAase. *See*: **restriction endonuclease**.

Dolly The first mammal (a sheep) to be created (via **nuclear transfer**) by the **cloning** of an adult **cell** (from the mammary tissue of a ewe). This showed that the process of differentiation into adult tissue is not, as previously thought, irreversible.

domain A portion of a **protein** or **DNA** molecule that has a discrete function or **conformation**. At the **protein** level, can be as small as a few **amino acid residues** or as large as half of the entire **protein**.

dominance The gene action exhibited by a **dominant allele**.

dominant (-acting) oncogene A gene that stimulates **cell proliferation** and contributes to **oncogenesis** when present in a single copy.

dominant 1. Of alleles, one whose effect with respect to a particular trait is the same in **heterozygotes** as in **homozygotes**. The opposite is **recessive**. 2. Of an individual animal, one that is allowed priority in access to food, mates, etc., by others of its species because of its success in previous aggressive encounters. 3. Of an animal or plant species, the most conspicuously abundant and characteristic in a particular location or environment.

dominant marker selection Selection of cells via a **gene** encoding a product that enables only the cells that carry the gene to grow under particular conditions. For example, plant and animal cells that express the introduced ***neo***^{r} gene are resistant to neomycin and analogous antibiotics, while cells that do not carry *neo*^{r} are killed. *See*: **positive selection.**

dominant selectable marker A gene that allows the host cell to survive under conditions where it would otherwise die. *Synonym*: **positive selectable marker**.

donor junction site The junction between the 5' end of an **exon** and the 3' end of an **intron**. *See*: **acceptor junction site**.

donor plant *See*: **ortet**.

dormancy A period in the life of an animal (hibernation and aestivation) or plant during which growth slows or completely ceases. Evolved to allow survival of adverse environmental conditions. Annual plants survive the winter as dormant seeds, while many perennial plants survive as dormant tubers, rhizomes, or bulbs. Premature breaking of seed dormancy post harvest can be a major problem for maintaining nutritional and/or functional quality, while difficulties in breaking dormancy will lead to poor **germination** of the crop. *See:* **quiescent**.

dosage compensation A regulatory mechanism for sex-linked genes, to allow equivalent levels of gene expression from (in mammals) XY or XX genotypes, even though the gene **copy number** in XX is double that in XY. *See:* **sex linkage, Barr body**.

double crossing-over The formation of two chiasmata within a **chromosome** arm, leading to the generation of a double **recombinant** gamete with respect to genes located within the segment defined by the two genes concerned.

double fertilization A process, unique to flowering plants, in which two male nuclei, which have travelled down the **pollen** tube, separately fuse with different female nuclei in the **embryo** sac. The first male **nucleus** fuses with the **egg cell** to form the **zygote**; the second male nucleus fuses with the two **polar nuclei** to form a **triploid** nucleus that develops into the endosperm.

double helix Describes the coiling of the two strands of **the double-stranded DNA** molecule, resembling a spiral staircase in which the **base pairs** form the steps and the sugar-phosphate backbones form the rails on each side. One strand runs 3'?5', while the complementary one runs 5'?3'

double recessive An organism homozygous for a **recessive allele** at each of two loci.

double-stranded complementary DNA (Abbreviation: dsc**DNA**). A double-stranded **DNA** molecule created from a **cDNA** template.

double-stranded DNA (Abbreviation: ds**DNA**). Two **complementary** strands of **DNA** annealed in the form of a **double helix**. *Synonym*: duplex **DNA**.

doubling time *See*: **cell generation time**.

down promoter mutation A mutation that decreases the frequency of initiation of **transcription**. This leads to a fall in the level of **mRNA** compared to the **wild type** state.

down-regulate To induce genetically a reduction in the level of a gene's expression.

downstream 1. With respect to **DNA**, the **nucleotides** that lie in the 3' direction from the point of reference, which is frequently the site at which **transcription** is initiated. This is generally designated +1, with downstream nucleotides numbered +2, +10 etc. 2. In chemical engineering, those phases of a manufacturing process that follow the **biotransformation** stage. Usually refers to the recovery and purification of the product of a **fermentation** process. *See*: **downstream processing**.

downstream processing A general term for biotechnological processes which follow the biology, i.e. fermentation of a **microorganism** or growth of a plant. Particularly relevant to **fermentation** processes, which produce a large quantity of a dilute mixture of substances, products and microorganisms. These must be separated, and the product concentrated, purified and converted into a useful form.

drift *See*: **genetic drift**.

Drosophila melanogaster The fruit fly, used for many years as a **model** for eukaryotic genetics. Of the nearly 300 disease-causing genes in the human genome, more than half have an analogous gene in the *Drosophila* genome.

drug delivery Method by which a drug is delivered to its site of action. For traditional drugs, this is another name for *formulation*. However, biotechnology has allowed the development of a range of new therapeutic-agent delivery systems, such as **liposomes** and other **encapsulation** techniques, and a range of mechanisms that target a **therapeutic** agent to a particular cell or tissue.

drug *See*: **therapeutic agent**.

dry weight The weight of **tissue** obtained following sufficiently prolonged oven-drying at high temperature to remove all water. **Freeze-**

drying may also be employed but generates a slightly different result because **bound water** is not removed. *See*: **free water**.

dscDNA *See*: **double-stranded complementary DNA.**

dsDNA *See*: **double-stranded DNA.**

dTTP Rarely used but strictly correct abbreviation for **deoxythymidine** 5'-triphosphate. Required for **DNA** synthesis since it is a direct precursor molecule. *See*: **TTP.**

dual culture A culture made of a plant tissue and one organism (such as a nematode) or an obligate parasite/microorganism (such as a fungus). Dual culture techniques are used for a variety of purposes, including assessing host-parasite interactions and the production of **axenic cultures**.

duplex DNA *See*: **double-stranded DNA.**

duplication Multiple occurrence of: 1. A **DNA sequence** within a defined length of **DNA**; or 2. A specific segment in the same **chromosome** or **genome**.

E site *See*: **exit site**.

E. coli *See*: ***Escherichia coli***.

EBV *See*: **estimated breeding value**.

EC *See*: **Enzyme Commission number**.

ecdysone A steroid **hormone** in insects stimulating the synthesis of **proteins** involved in moulting and metamorphosis.

eclosion 1. Emergence of an adult insect from the pupal stage. 2. Initial phase of germination of fungal spores.

ecological diversity *See*: **biodiversity**.

economic trait locus (Abbreviation: ETL). A locus influencing a trait that contributes to producer's income.

ecosystem The complex of a living community and its environment, functioning as an ecological unit in nature. *See:* **abiotic**; **biotic factors**.

ecotype A population or a strain of an organism that is adapted to a particular habitat.

ectopic Anomalous situation or relation, particularly with respect to pregnancy, where the **foetus** is implanted outside the uterus.

edible vaccine Edible antigen-containing material, that activates the immune system via gut-associated lymphoid tissues. A preferred route for vaccine administration, particularly in areas where the technological infrastructure needed for maintenance of vaccines is absent. The vaccine is synthesized *in vivo* in the edible parts of **transgenic** plants (e.g. grains, tubers, fruits, etc.) or eggs.

editing *See*: **splicing** (1).

EDTA *See*: **ethylenediamine tetraacetic acid**.

EDV Abbreviation for **essential derivation of varieties**.

effector cells Cells of the immune system that are responsible for the production of cell-mediated **cytotoxicity**.

effector molecule A molecule that influences the behaviour of a regulatory molecule, such as a **repressor protein**, thereby influencing **gene** expression.

egg 1. The fertilized **zygote** in egg-laying animals. 2. The mature female reproductive cell in animals and plants.

EGS *See*: **external guide sequence**.

EIA *See*: **enzyme immunoassay, ELISA**.

elastin A fibrous **protein** that is the major constituent of the yellow elastic fibres of animal connective tissue.

electro-blotting The electrophoretic transfer of **DNA, RNA** or **protein** from a gel, in which they have been separated, to a support matrix, such as **nitrocellulose**. A transfer technique employed in **Southern** and **northern** blotting.

electrochemical sensor Biosensors, such as an **enzyme electrode**, in which a biological process is harnessed to an electrical sensor system. Other types couple a biological event to an electrical one via a range of mechanisms, including the reduction of oxygen or **pH** change.

electron microscope (Abbreviation: EM). A microscope that uses an electron beam focussed by magnetic 'lenses'. *See*: **scanning electron microscope**.

electrophoresis A ubiquitous molecular biology

technique, with many variants, used to resolve complex mixtures of macromolecules into their components. Its principle is to subject samples to an electric field applied across a porous matrix. Molecules will migrate under these conditions at a rate dependent on their net electric charge and/or their molecular weight. *See*: **agarose gel electrophoresis, polyacrylamide gel electrophoresis, denaturing gradient gel electrophoresis, capillary electrophoresis, sodium dodecyl sulphate polyacrylamide gel electrophoresis, thermal gel gradient electrophoresis pulsed-field gel electrophoresis,** and **iso-electric focusing gel.**

electroporation The induction of transient pores in bacterial cells or protoplasts by the application of a pulse of electricity. These pores allow the entry of **exogenous DNA** into the cell. Widely used for the **transformation** of bacteria.

ELISA Abbreviation for enzyme-linked immunosorbent assay. An **immunoassay**, i.e. an **antibody**-based technique for the diagnosis of the presence and quantity of specific molecules in a mixed sample. It combines the specificity of an **immunoglobulin** with the detectability of an enzyme-generated coloured product. In one form, the primary **antibody** (specific to the test **protein**) is adsorbed onto a solid substrate, and a known amount of the sample is added; all the **antigen** in the sample is bound by the antibody. A second antibody (conjugated with an enzyme) specific for a second site on the test protein is added; and the **enzyme** generates a colour change in the presence of a substrate reagent.

elite tree A phenotypically superior tree in a tree **breeding** programme.

elongation factors Soluble **proteins** required for the elongation of **polypeptide** chains on **ribosomes.**

embryo An immature organism in the early stages of development. In mammals, develops in the first months in the uterus. In plants, it is the structure that develops in the **megaga-**

metophyte, as result of the **fertilization** of an **egg** cell, or occasionally without fertilization. **Somatic embryos** can often be induced in *in vitro* plant cell cultures.

embryo cloning The creation of identical copies of an **embryo** by **embryo splitting** or by **nuclear transfer** from undifferentiated embryonic cells.

embryo culture The culture of embryos on nutrient media.

embryo multiplication and transfer (Abbreviation: EMT). The cloning of animal embryos and their subsequent transfer to recipients via **artificial inembryonation**. The cloned embryos can be derived from embryonic or adult tissue.

embryo rescue A sequence of **tissue culture** techniques utilized to enable a fertilized immature embryo resulting from an **interspecific cross** to continue growth and development, until it can be regenerated into an adult plant.

embryo sac The mature female **gametophyte** in angiosperms. Generally a seven-celled structure - two synergids, one **egg** cell, three antipodal cells (each with a single haploid nucleus) and one **endosperm** mother cell with two **haploid** nuclei.

embryo sexing The determination of the sex of an embryo prior to birth. Typically achieved by the **polymerase chain reaction**-mediated amplification of **DNA** extracted from a sample of embryonic tissue. Dependent on the availability of reliable markers for the differential sex chromosome.

embryo splitting The splitting of young embryos into several sections, each of which develops into an animal. A form of animal cloning, i.e. of producing animals that are genetically identical. In practice, the number of animals that can be produced from a single embryo is less than 10.

embryo storage Cryogenic preservation of animal embryos, allowing **inembryonation** or other manipulations long after embryo formation.

embryo technology Generic name for any modification of mammalian embryos. It encompasses **embryo cloning, embryo splitting, embryo storage**, and ***in vitro* fertilization.**

embryo transfer (Abbreviation: ET). *See*: **embryo multiplication and transfer, multiple ovulation and embryo transfer.**

embryogenesis 1. (General) Development of an embryo. 2. (In plants) *In vitro* formation of plants from plant tissues, through a pathway closely resembling normal embryogeny from the zygote. **Somatic cell embryogenesis** is an alternative technique. The generation of embryos has two stages: initiation and maturation. Initiation needs a high level of the group of plant hormones called auxins; maturation needs a lower level.

embryoid Plant biotechnology term no longer commonly used. An embryo-like body developing *in vitro*, forming a complete, self-contained plantlet with no **vascular** connection with the **callus.**

embryonic stem cells (Abbreviation: ES cells). Cells of the early embryo that can give rise to all differentiated cells, including **germ line** cells.

emission wavelength The specific wavelength of light emitted by a fluorescent molecule, such as a labelled **probe,** upon absorption of light at the (higher) **excitation wavelength.**

EMT *See*: **embryo multiplication and transfer.**

encapsidation The process by which the **nucleic acid** of a **virus** is enclosed in a **capsid.**

encapsulating agents Anything which forms a shell around an **enzyme** or bacterium, common agents being **polysaccharides** such as **alginate** or agar. The agents are inert and allow nutrients and oxygen to diffuse readily into and out of the sphere, and are easy to convert from gel (solid) to sol (liquid) or solution form by altering the temperature or the concentration of ions.

encapsulation Any method packaging an **enzyme** or bacterium and maintaining its normal functions. Used to immobilize cells in a bioreactor.

encode The gene product specified by a particular **nucleic acid** sequence. *See*: **genetic code**.

endangered species A plant or animal species in immediate danger of extinction because its **population** number has reached a critical level, or its habitat has been drastically reduced.

endemic Describing an organism, often a disease or pest, that is always present in a stated area.

end-labelling The introduction of a readily-visualized **tag** at the end of a **DNA** or **RNA** molecule. A commonly used method is to introduce a ^{32}P atom onto the end of a **DNA** molecule by means of the enzyme T4 **polynucleotide** kinase.

endocrine gland Any **gland** in an animal that manufactures hormones and secretes them directly into the bloodstream to act at distant sites in the body, known as target organs or cells.

endocrine interference Interference with the normal balance of hormones.

endocytosis The process by which materials enter a **cell** without passing through the cell membrane. The membrane folds around material outside the cell, resulting in the formation of a sac-like vesicle inside which the material is entrapped. This vesicle is then pinched off from the cell surface so that it lies within the cell. *See*: **phagocytosis, pinocytosis**.

endoderm The internal layer of cells of the **gastrula**, which develops into the alimentary canal (gut) and digestive glands of the adult.

endodermis The layer of living cells, with various characteristically thickened walls and no intercellular spaces, which surrounds the **vascular tissue** of certain plants and occurs in nearly all roots and certain stems and leaves. It separates the cortical cells from cells of the **pericycle**.

endogamy *See*: **inbreeding**.

endogenous Derived from within; from the same **cell** type or organism. *Opposite*: **exogenous**.

endomitosis Duplication of chromosomes without division of the nucleus, resulting in a doubling (or more) in the **chromosome** number within a cell.

endonuclease An enzyme that cleaves a **phosphodiester bond** within a **DNA** strand, forming two smaller strands. *See*: **exonuclease, restriction endonuclease.**

endophyte An organism that lives inside a plant.

endoplasmic reticulum (Abbreviation: ER). A cytoplasmic net of membranes, adjacent to the nucleus, visible under the electron microscope. The sites of **protein** synthesis.

endopolyploidy The net result of **endomitoses**. The somatic **chromosome** number has doubled (or more), forming a **polyploid** cell line. If these differentiate into a **germ line**, then the gametic number will have also increased proportionately, giving rise to homogeneously polyploid individuals, termed endopolyploids.

endoprotease An enzyme that cleaves internal **peptide** bonds within a **polypeptide** molecule. Site of cleavage is usually specific to certain amino acid **residues**.

endoreduplication Chromosome reproduction during interphase. Four-chromatid chromosomes (diplochromosomes) are seen during this phase.

endosperm mother cell One of the seven cells of the mature plant **embryo** sac, containing the two **polar nuclei** and, which, following fertilization, gives rise to the primary endosperm cell from which the **endosperm** develops.

endosperm The nutritive tissue that develops in the seed of most angiosperms, containing varying proportions of **carbohydrate** (usually starch), **protein** and lipid. In most **diploid** plants, the endosperm is **triploid.**

endotoxin A component of the cell wall of gram-negative bacteria that elicits, in mammals, an inflammatory response and fever.

end-product inhibition The inhibition of an **enzyme** by a **metabolite**. Typically, the enzyme is the first enzyme in a biosynthetic pathway, and the metabolite the product of the last step in the pathway. *See*: **feedback inhibition**.

enhancer 1. A substance or object that increases a chemical activity or a physiological process. 2. A eukaryotic **DNA** sequence (also found in some eukaryotic viruses) which increases the **transcription** of a gene. Located up to several kbp, usually (but not exclusively) **upstream** of the gene in question. In some cases can activate transcription of a gene with no (known) **promoter**. *Synonyms*: **enhancer element; enhancer sequence**. 3. A major or modifier gene that increases the rate of a physiological process.

enhancer element *See*: **enhancer**.

enhancer sequence *See*: **enhancer**.

enolpyruvyl-shikimate-3-phosphate synthase (Abbreviation EPSP synthase or EPSPS). An enzyme produced by virtually all plants, which is essential for normal metabolism, and for the **biosynthesis** of aromatic amino acids. **Glyphosate**- and sulfosate-containing herbicides act by inhibiting **EPSP synthase** activity, but because strain CP4 of ***Agrobacterium*** sp. is unaffected by glyphosate, the introduction of the ***CP4 EPSPS*** gene into crop plants generates a tolerance of glyphosate-containing herbicides.

enterotoxin A bacterial **protein** that, following release into the intestine, causes cramps, diarrhoea and nausea.

enucleated ovum Egg cell from which the **nucleus** has been removed, usually as a preparatory step for **nuclear transfer**.

enzyme A **protein** which, even in very low concentration, catalyses specific chemical reactions but is not used up in the reaction. Enzymes are classified into six major groups (1-6), according to the type of reaction they catalyse: 1. oxidoreductases; 2. transferases; 3. hydrolases; 4. lyases; 5. isomerases; 6. ligases. Generally, enzymes are named by the addition of the suffix -ase to the name of their **substrate**, and are classified by a standard numerical system: the **Enzyme Commission (EC) number**.

enzyme bioreactor A reactor in which a chemical **conversion** reaction is catalysed by an **enzyme.**

Enzyme Commission number (Abbreviation: EC number). Systematic **label** which identifies an enzyme in the technical literature. Consists of four numbers separated by dots: the first classifies the enzyme into one of the six broad enzyme groups (*see*: **enzymes**); each group is subdivided into sub-groups, each sub-group into sub-sub-groups, and the last number is specific for the enzyme, e.g. EC 3.1.21.1 is **deoxyribonuclease** I.

enzyme electrode A type of biosensor, in which an **enzyme** is immobilized onto the surface of an electrode. When the enzyme catalyses its reaction, electrons are transferred from the reactant to the electrode, and so a current is generated. There are two types of enzyme electrodes: 1. Ampometric (measuring current flow) where the electrode is kept as near zero voltage as possible. When the enzyme catalyses its reaction, electrons move into the electrode, and so a current flows; 2. **Potentiometric** (measuring changes in electrical potential) when the electrode is held at a voltage which counteracts the voltage determined by the enzyme's tendency to push electrons into it. Usually, enzymes transfer their electrons inefficiently to the electrode, so a mediator compound is coated onto the electrode to help the transfer.

enzyme immunoassay A range of immunoassay techniques employing enzymes, which includes **ELISA**.

enzyme kinetics The quantitative characteristics of enzyme reactions.

enzyme stabilization Maintaining the active **conformation** of an **enzyme**. This can be achieved *in vitro* by providing the appropriate chemical environment and cofactors. In some cases, the criticality of these factors can be reduced by binding an **antibody** to the enzyme, in such a way that the active site of the enzyme is left unblocked.

enzyme-linked immunosorbent assay *See*:

ELISA.

EPD *See*: **expected progeny difference**.

epicotyl The upper portion of the axis of a plant **embryo** or seedling, above the **cotyledons**.

epidermis 1. The outermost layer of cells of the body of an animal. In invertebrates the epidermis is normally only one **cell** thick and is covered by an impermeable **cuticle**. In vertebrates the epidermis is the thinner of the two layers of skin. 2. The outermost layer of cells covering a plant. It is overlaid by a cuticle and its functions are principally to protect the plant from injury and to reduce water loss. Some epidermal cells are modified to form **guard cells** or hairs of various types. In woody plants, the functions of the shoot epidermis are taken over by the periderm tissues and in mature roots the epidermis is sloughed off and replaced by the hypodermis.

epigenesis Describes the developmental process whereby each successive stage of normal development is built up on the foundations created by the preceding stages of **development**; an embryo is built up from a zygote, a seedling from an **embryo**, and so on.

epigenetic variation Non-hereditary and reversible variation; often the result of a change in **gene expression** due to **methylation** of **DNA**.

epinasty A process by which the growth of branches or petioles is abnormally pointing downward. This phenomenon is caused by the more rapid growth of the upper side. Epinasty may result from either nutritional deficiencies or irregularities at the plant **growth regulator** level. Not to be confused with wilting, as epinastic tissues are turgid.

epiphyte A plant that grows upon another plant, but is neither parasitic on it nor rooted in the ground.

episome A genetic **extrachromosomal** element (e.g. the **F factor** in *Escherichia coli*) which replicates within a cell independently of the chromosome and is able to integrate into the **host** chromosome. The step of integration may be governed by a variety of factors and so the term episome has lost favour and been superseded by the wider term **plasmid**.

epistasis Interaction between genes at different loci, e.g. one **gene** suppresses the effect of another gene that is situated at a different locus. **Dominance** is associated with members of **allelic** pairs, whereas epistasis describes an interaction among products of non-alleles.

epitope Synonym for **antigenic determinant**.

epizootic A disease simultaneously affecting a large number of animals.

EPSP synthase Abbreviation for **enolpyruvylshikimate 3-phosphate synthase**.

EPSPS Abbreviation for **enolpyruvylshikimate 3-phosphate synthase**.

equational division A chromosome division in which the two chromatids of each duplicated chromosome separate longitudinally, prior to being incorporated into two daughter nuclei. Seen at the mitotic-type second division of **meiosis**; also in somatic **mitosis** and the non-reductional division of meiosis. The number of chromosomes is the same at the end of the division as at the beginning.

equilibrium density gradient centrifugation A procedure used to separate macromolecules based on their density (mass per unit volume).

ER *See*: **endoplasmic reticulum.**

Erlenmeyer flask A conical flat-bottomed laboratory flask with a narrow neck, widely used for culturing microorganisms.

ES cells *See*: **embryonic stem cells.**

Escherichia coli A commensal bacterium inhabiting the colon of many animal species, including human. *E. coli* is widely used as a **model** of cell biochemical function, and as a host for **cloning DNA**. In environmental studies, its presence is a key indicator of water pollution due to human sewage effluent. Some strains, notably *E. coli* 0157:H7, are significant **pathogens**.

essential amino acid An **amino acid** required for normal metabolism, but which cannot be synthesized by an organism. It therefore has to be supplied via feed or food.

essential derivation of varieties (Abbreviation: EDV). Genotypes very similar to an originating **cultivar,** obtained, for example, by the **selection** of a **mutant** or a variant individual from plants of the initial variety, or by backcrossing or **transformation**.

essential element Any of a number of elements required by living organisms to ensure normal growth, development and maintenance.

essential nutrient Any substance required by living organisms to ensure normal growth, development and maintenance.

essential requirement In plant cell **tissue culture**, comprises inorganic salts, including all of the elements necessary for plant metabolism; organic factors (amino acids, vitamins); usually also endogenous plant growth regulators (auxins, cytokinins and often gibberellins); as well as a carbon source (sucrose or glucose).

EST *See*: **expressed sequence tag**.

established culture 1. An aseptic viable **explant** (*See*: **micropropagation**). 2. A **suspension culture** subjected to several **passages** with a constant cell number per unit time.

estimated breeding value (Abbreviation: EBV). Twice the **expected progeny difference**. The difference is doubled because breeding value is a reflection of all the genes of an individual, in contrast to **progeny** difference, which is a reflection of a sample half of an individual's genes. The predicted performance of the **offspring** of the mating between any two parents is the average of their EBVs (averaged because each parent makes an equal contribution to each offspring).

estrogen *See*: **oestrogen**.

ET Abbreviation for **embryo transfer**. *See*: **multiple ovulation and embryo transfer**.

ethanol Commonly used to disinfect plant tissues, glassware utensils and working surfaces in tissue culture manipulations; to precipitate aqueous solutions of nucleic acids; and to dissolve water-insoluble components of culture media. *Synonym*: **ethyl alcohol**.

ethephon A synthetic compound commonly

used as a source of **ethylene**, a gaseous plant growth regulator.

ethidium bromide A fluorescent dye which can intercalate between base pairs of **double-stranded DNA**, and hence is much used to stain **DNA** in gels. The dye fluoresces when exposed to UV light. It is a known to be a strong **mutagen** and is also possibly both a **carcinogen** and a teratogen.

ethyl alcohol *See*: **ethanol.**

ethylene A gaseous plant **growth regulator** acting on various aspects of vegetative growth, fruit ripening and abscission of plant parts. *Synonym*: ethene.

ethylenediamine tetraacetic acid (Abbreviation: EDTA). A chelating compound. Used to keep nutrients, such as iron, bound in a soluble form that leaves them still available to the plant cells *in vitro*. Also a potent **inhibitor** of **DNase** activity and therefore used as an additive for long-term storage of dissolved **DNA.**

etiolation An abnormal increase in **stem** elongation, accompanied by poor (if any) leaf development. Physiological etiolation is caused by a lack of chlorophyll, and is typical of plants growing under low light intensity or in complete darkness. It can also be induced by some fungal pathogens.

ETL *See*: **economic trait locus.**

eucaryote *See*: **eukaryote.**

eucaryotic (adj.) *See*: **eukaryote.**

euchromatin Chromosomal material that is stained less intensely by certain dyes. Thought to be the chromosomal domains which are gene-rich, since the **DNA** in these regions remains less contracted than those rich in **repetitive DNA** - the **heterochromatin.**

eugenics The application of the principles of genetics to the 'improvement' of humankind. Wholly discredited as a scientific approach since the Nazi period.

eukaryote One of the two major evolutionary clades, characterized by having the **nucleus** enclosed by a membrane, and possessing chromosomes that undergo **mitosis** and **meiosis**. Eukaryotic organisms include animals, plants, fungi and some algae. *See*: **prokaryote**.

euploid An organism or cell having a **chromosome** number that is an exact multiple of the haploid number. Terms used to identify different levels in an euploid series are **diploid** (2x), **triploid** (3x), **tetraploid** (4x) etc. *Opposite*: **aneuploid**.

evapotranspiration The net water loss (in vapour form) per unit area of land, both directly from the land surface, and indirectly through transpiring leaves.

evolution The process by which the present diversity of plant and animal life has arisen, and which continues to drive changes in form and mode of existence of all living organisms.

ex vitro Organisms removed from **tissue culture** and transplanted; generally plants to soil or potting mixture.

***ex vivo* gene therapy** The delivery of a gene or genes to the isolated cells of an individual, with the intention of alleviating a genetic disorder. After culturing, the transformed cells are re-introduced into the individual by transfusion, infusion or injection.

excinuclease The **endonuclease**-containing **protein** complex that excises a segment of damaged **DNA** during **excision** repair.

excision 1. The natural or *in vitro* enzymatic removal of a **DNA** segment from a **chromosome** or **cloning** vector. 2. The cutting out and preparation of a tissue, organ, etc., for culture. 3. The removal of adventitious shoots from callus tissue.

excision repair DNA repair processes that involve the removal of a damaged or incorrect segment of one strand of **double-stranded DNA** and its replacement by the synthesis of a new segment using the **complementary** strand of **DNA** as template.

excitation wavelength The specific wavelength of light required to stimulate a fluorescent molecule, such as a labelled **probe**, to emit light at the (lower) **emission wavelength**.

excrete To transport material out of a **cell** or

organism.

exit site (Abbreviation: E site). The **ribosome** binding site that contains the free **tRNA** prior to its release.

exo III *See*: **exonuclease III.**

exocrine gland An animal **gland** that secretes through a duct.

exodeoxyribonuclease III *See*: **exonuclease III.**

exogamy *See*: **outbreeding.**

exogenous DNA DNA that has been derived from one organism, and is to be introduced into a cell a different species. Also referred to as **foreign DNA** or **heterologous DNA.**

exogenous Produced outside of; originating from, or due to, external causes. *Opposite*: **endogenous.**

exon A segment of a eukaryotic gene that is transcribed as part of the primary **transcript** and is retained, after processing, with other exons to form a functional **mRNA** molecule. Many eukaryotic genes are composed of a **mosaic** of exons and **introns.**

exon amplification A procedure that is used to **amplify** exons.

exonuclease An enzyme that digests **DNA** or **RNA**, beginning at the end of a strand. It therefore requires a free end in order to begin the degradation. 5'-exonucleases require a free 5' end and degrade the molecule in the 5'?3' direction. 3'-exonucleases require a free 3' end and degrade in the opposite direction.

exonuclease III (Abbreviation: exo III). An ***Escherichia coli*** enzyme that removes nucleotides from the 3' hydroxyl ends of double-stranded **DNA.** *Synonym*: exodeoxyribonuclease III.

exopolysaccharide A **polysaccharide** that is secreted by a **microorganism** into the surrounding environment.

exotoxin A **toxin** released by a bacterium into the medium in which it grows.

expected progeny difference (Abbreviation: EPD). The predicted performance of the future **offspring** of an individual for a particular **trait**, calculated from measurement(s) of the individual's own performance and/or the performance of one or more of its relatives, for the trait in question and/or for one or more correlated traits. Typically, the prediction is expressed as a **deviation** from a well-defined base population, assuming the individual in question is mated to a sample of individuals whose average genetic merit equals that of the base population. The predicted performance of the offspring of the mating between any two individuals is the sum of their EPDs.

explant A portion of a plant aseptically excised and prepared for **culture** in a nutrient medium.

explant donor The plant from which an **explant** has been taken.

explantation The removal of cells, tissues or organs of animals and plants for observation of their growth and development in appropriate **culture** media.

explosion method A technique for the genetic **transformation** of cells, in which the **transgene** is driven into the target (plant) cells by the sudden vaporization (effected by the application of a pulse of high voltage) of a water droplet containing the **DNA** and gold particles.

exponential phase *See*: **logarithmic phase**.

export The removal of a compound from a cell by **active transport**.

express To transcribe and translate a gene.

expressed sequence tag (Abbreviation: EST). Partially sequenced **cDNA** clone. Because the read length of a standard **DNA sequencing** reaction is shorter than the majority of **cDNA** clones, full length sequence can only be obtained by further manipulations. For the purposes of (1) assigning putative function to a cDNA and (2) designing **PCR primers** to extract the genomic **DNA** equivalent to the cDNA, full length sequence is usually unnecessary. By restricting sequencing to a single run, large numbers of cDNAs can be characterized at the EST level.

expression library A **cDNA library** that has been inserted into a bacterial host cell engineered to **express transgenes**. *See*: **library**.

expression system Combination of **host** and **vector** which provides a genetic context for making a cloned gene functional, i.e. produce **peptide**, in the host cell.

expression vector A cloning **vector** that has been constructed in such a way that, after insertion of a **DNA** molecule, its **coding** sequence is properly transcribed and the **mRNA** is translated. The cloned gene is put under the control of a **promoter** sequence for the initiation of transcription, and often also has a **transcription** termination sequence at its end.

expressivity Degree of expression of a **trait** controlled by a particular **gene**. The gene may show different degrees of expression in different individuals. *See*: **variable expressivity**.

***ex-situ* conservation** The conservation of components of biological diversity outside their natural habitats.

extension The short single-stranded stretch of nucleotides remaining on a **double-stranded DNA** molecule, following treatment with a **restriction endonuclease** which makes a **staggered cut**. The presence of these unpaired regions make the molecule more easily ligatable, and are thus important in gene **cloning**. *Synonyms*: **protruding end**; **sticky end**; **overhang**; **cohesive end**.

external guide sequence (Abbreviation: EGS). *See*: **guide sequence**.

extrachromosomal In eukaryotes, non-nuclear **DNA**, present in **cytoplasm** organelles such as mitochondria and **chloroplasts.** In prokaryotes, non-chromosomal **DNA**, i.e. **plasmids**.

extrachromosomal inheritance *See*: **cytoplasmic inheritance**.

extranuclear genes Genes residing elsewhere than in the **nucleus** (e.g. in mitochondria, chloroplasts or plastids).

exude Slowly discharge liquid material (such as tannins or oxidized polyphenols from plant material) through pores or cuts, or by **diffusion** into the medium.

F

F factor Abbreviation for fertility factor. A bacterial **plasmid** that confers the ability to function as a genetic donor in **conjugation**. *See*: **Hfr**.

$\mathbf{F_1}$ Abbreviation for **filial generation** 1. The initial **hybrid** generation resulting from a **cross** between two parents. *See* $\mathbf{F_n}$.

$\mathbf{F_1}$, $\mathbf{F_2}$, $\mathbf{F_n}$ Subsequent **hybrid** generations, counting from the $\mathbf{F_1}$. Thus, for example, F_4 describes the **progeny** of the F_3, which is the progeny of the $\mathbf{F_2}$ generation, where all progeny are derived from intercrossing or self-fertilization.

$\mathbf{F_2}$ The second **hybrid** generation, produced either by intercrossing two $\mathbf{F_1}$ individuals, or by self-fertilizing an F_1 individual. *See* $\mathbf{F_n}$.

$\mathbf{F_{ab}}$ A product of **hydrolysis** of an IgG **antibody**, consisting of the variable region with some of the constant region of a heavy chain, and an entire light chain. Contains a single **antigen**-binding site.

FACS *See*: **fluorescence-activated cell sorting**.

factorial mating A mating scheme in which each

male parent is mated with each female parent. Made possible in animals by means of ***in vitro* embryo production**. Such a mating scheme substantially reduces the rate of **inbreeding** in a **selection** programme.

facultative anaerobe An organism that will grow under either **aerobic** or **anaerobic** conditions.

FAD *See*: **flavin adenine dinucleotide**.

false fruit *See*: **pseudocarp**.

false negative A negative **assay** result that should have been positive.

false positive A positive **assay** result that should have been negative.

farm animal genetic resources Those animal species that are used, or may be used, for the production of food and agriculture, and the populations within each of them. Within each species, these populations can be classified as wild and feral populations, landraces and primary populations, standardized breeds, selected lines, and any conserved genetic material.

farmers' privilege Rights to hold germplasm, covered by **plant variety protection,** as a seed source for subsequent seasons. Considered as optional for governments to include in their legislation. *Synonym*: farmer-saved seed.

farmers' rights Rights first recognized by Resolution 5 of the 1989 FAO Conference as "rights arising from the past, present and future contributions of farmers in the conservation, improvement and the making available of **plant genetic resources**"; this item became an attachment to the **'International Undertaking on Plant Genetic Resources'**. The binding **'International Treaty of Plant Genetic Resources for Food and Agriculture'** that resulted from the renegotiations of the Undertaking makes provision for the Farmers' Rights in Article 9.

fascicle *See:* **vascular bundle**.

F_c A product of **hydrolysis** of an IgG **antibody**, consisting of parts of the constant regions of two heavy chains held together by a **disulphide bridge**, but excluding the **anti-**

gen-binding regions, and also excluding the light chains.

fed-batch fermentation Culture of cells or microorganisms where nutrients are added periodically to the bioreactor.

feedback inhibition The process by which the accumulated end product of a biochemical pathway stops synthesis of that product. The effect is that a late **metabolite** of a synthetic pathway regulates the synthesis of an earlier step of the pathway. *See*: **end-product inhibition**.

fermentation substrates Materials used as food for growing microorganisms. The fermentation substrates and the trace materials needed, together with chemicals added to make the fermentation easier, form the **culture** medium.

fermentation The **anaerobic** breakdown of complex organic substances, especially carbohydrates, by microorganisms, yielding energy. Often misused to describe large-scale **aerobic cell culture** in specialized vessels (fermenters, bioreactors) for secondary product synthesis.

fermenter *See*: **bioreactor**.

fertile Capable of **breeding** and reproduction.

fertility factor *See*: **F factor**.

fertilization The union of two gametes from opposite sexes to form a zygote. Typically, each **gamete** contains a **haploid** set of chromosomes. Hence, the zygotic **nucleus** contains a **diploid** set of **chromosomes**. Several categories can be distinguished: 1. Self-fertilization (selfing): fusion of male and female gametes from the same individual. 2. Cross-fertilization (crossing): fusion of male and female gametes from different individuals. 3. Double fertilization; restricted to flowering plants, in which the fusion of one male **gamete** with the **ovum** occurs at about the same time as the second male gamete nucleus fuses with the female **polar nuclei** (or secondary nucleus) to form the **endosperm**.

fertilizer Any substance that is added to soil in order to increase its productivity. Fertilizers can be of biological origin (e.g. composts), or they can be synthetic (artificial fertilizer).

fetus *See*: **foetus.**

Feulgen staining A histochemical stain by which the distribution of **DNA** in the **chromosomes** of dividing **cell** nuclei can be observed.

FIA Abbreviation for **fluorescence immunoassay.**

fibril A microscopic to sub-microscopic cellulose thread that is part of the **cellulose** matrix of plant **cell** walls.

fibroblasts Irregularly shaped, branching cells distributed throughout vertebrate connective tissue. A **cell** type which is readily cultured ***in vitro*.**

fibrous root Root system in which both primary and lateral roots have approximately equal diameters. *Opposite*: **taproot.**

field gene bank *See*: **gene bank** (2).

filial generation *See*: F_1, F_2, F_n.

filter bioreactor A cell culture system, in which cells are grown on a fine mesh of an inert material, which allows the culture medium to flow past it but retains the cells. This is similar in idea to membrane and **hollow fibre** reactors, but can be much easier to set up, being similar to conventional tower bioreactors, but with the mesh replacing the central reactor space. *Synonym*: **mesh bioreactor.**

filter sterilization Process of removing microbial contaminants from a liquid by passing through a filter with pores too small to allow the passage of **microorganisms** and spores.

filtration 1. Separation of solids from liquids by using a porous material that only allows passage of the liquid or of solids smaller than the pore size of the filter. The material passing the filter forms the filtrate. 2. Removal of cell aggregates to obtain a filtrate of single cells that can be utilized as plating inocula.

fingerprinting *See*: **DNA fingerprinting.**

FISH *See*: **fluorescence *in situ* hybridization**.

fission Asexual reproduction involving the division of a single-celled individual into two daughter single-celled individuals of approximately equal size.

fitness The survival value and the reproductive capability of an individual, compared to that of competitor individuals of the same or other species within a **population** or an environment.

fixation The situation in which only one allele for a given gene/locus is present in a population. This can occur as a result of direct selection where the allele delivers a greater level of fitness; because of indirect selection, where the locus is linked to a gene that is subject to direct selection; or because of **genetic drift**.

FLAG *See* affinity tag.

flaming A technique for sterilizing instruments, to remove **live microorganism** contaminants. The instrument is dipped in alcohol, and the alcohol remaining on the instrument is ignited, thereby heat-sterilizing the surface.

flanking region The **DNA** sequences extending either side of a specific sequence.

flavin adenine dinucleotide (Abbreviation: FAD). A **co-enzyme** important in various biochemical reactions. It comprises a phosphorylated **vitamin** B_2 (riboflavin) molecule linked to **AMP**, and functions as a hydrogen acceptor in **dehydrogenation** reactions. The reduced form is oxidized back to FAD by the electron transport chain, generating two molecules of **ATP** per molecule of reduced FADH.

flocculant A chemical agent that causes small particles to aggregate (flocculate).

floccule A microorganism aggregate or colloidal particle floating in or on a liquid. The cloudy appearance of **microorganism** contaminated liquid media illustrates the flocculation phenomenon.

flow cytometry Automated measurements on large numbers of individual cells or other small biological materials, made as the cells flow one by one in a fluid stream past optical and/or electronic sensors. A similar approach may be used for sorting cells - see **fluorescence-activated cell sorting**.

fluorescence immunoassay (Abbreviation: FIA).

An **immunoassay** based on the use of fluorescence-labelled antibody.

fluorescence *in situ* hybridization (Abbreviation: FISH). **Hybridization** of cloned, fluorescently labelled **DNA** or **RNA**, to intact biological materials, notably **chromosome** spreads and thin tissue sections. The technique allows the visualization of the physical location of **nucleic acid** sequences **homologous** to the **probe**, and is used for the placement of genes on chromosomes and for the spatial and temporal pattern of **gene expression** of specific **mRNA** molecules.

fluorescence-activated cell sorting (Abbreviation: FACS). A **flow cytometry** method in which targets (cells, individual chromosomes etc.) are labelled with a fluorescent dye, which is excited by a laser beam. Differences in the fluorescence signal emitted are used as a criterion for sorting the material. A specific application is in **sperm sexing**.

fluorescent probe A **probe** which is labelled with a fluorescent dye, so that the signal emitted can be captured by photometric methods.

flush end *See*: **blunt end**.

flush-end cut *See*: **blunt-end cut**.

foetus Pre-natal stage of a **viviparous** animal, between the embryonic stage and birth. *Alternative spelling*: fetus. *See*: **embryo**.

fog Fine particles of liquid suspended in the air, such as of water in a fog chamber used for acclimatizing recent *ex vitro* transplants. *See*: **mist propagation**.

fold-back The structure of a **double-stranded DNA** molecule formed when a molecule containing an **inverted repeat** sequence is denatured and then allowed to re-**anneal** at low **DNA** concentrations. Under these conditions, the repeated sequence self-anneals to form a double-stranded region within each of the separated strands of the original molecule.

folded genome The condensed state of the chromosomal **DNA** of a bacterium. The **DNA** is segregated into **domains**, and each domain is independently negatively supercoiled.

follicle An enclosing cluster of cells that protects and nourishes a **cell** or structure within. Thus, a follicle in the **ovary** contains a developing **egg** cell, while a hair follicle envelops the root of hair.

follicle stimulating hormone (Abbreviation: FSH). A hormone, secreted by the anterior pituitary **gland** in mammals, that stimulates the ripening of the specialized structures in the **ovary** (Graafian follicles) that produce ova in female mammals; and in males, the formation of **sperm** in the testis. FSH is a major constituent of fertility drugs.

food processing enzyme Enzyme used to control food texture, flavour, appearance, or nutritional value. Amylases break down complex polysaccharides to simpler sugars; proteases tenderize meat **proteins**. A prominent target of food **biotechnology** is to develop novel food enzymes which can improve the quality of processed foods.

forced cloning The insertion of **foreign DNA** into a cloning **vector** in a predetermined orientation.

foreign DNA **Exogenous DNA** that is incorporated into a **host** genome.

formulation *See*: **medium formulation**.

forskolin A medicinal, diterpenoid, compound exclusive to plant roots and used in the preparation of drugs for the treatment of cardiomyopathy, glaucoma and certain cancers.

fortify To add strengthening components or beneficial ingredients to a nutrient medium.

forward mutation A mutation from the **wild type** to the mutant type. *Opposite*: **reverse mutation**.

fouling The coating or plugging (by materials or microorganisms) of equipment, thus preventing it from functioning properly.

founder animal An organism that carries a **transgene** in its **germ line** and can be used in matings to establish a pure-breeding **transgenic** line, or one that acts as a breeding stock for transgenic animals.

founder principle The possibility that a new, isolated **population**, initiated by a small number of individuals taken from a parent population, may be genetically different from the parent population, because the founding individuals might not be typical of the parent population. *See*: **genetic drift.**

four-base cutter A type II **restriction endonuclease** with a four-nucleotide recognition site. Because any particular sequence of four bases occurs more frequently by chance than one of six bases, four-base cutters **cleave** more frequently than **six-base cutters**, and therefore generate, on average, smaller **restriction fragments**. *Synonyms*: four-base-pair-cutter, four-cutter.

fractionation The separation in components of a complex mixture of molecules.

fragment Partial structure. *See*: **restriction fragment.**

frameshift mutation A mutation that changes the **reading frame** of a **DNA**, either by the insertion or the **deletion** of nucleotides. Because of the **triplet** nature of **codons**, this occurs if the number of nucleotides involved is not a multiple of three.

free water The cellular water released into the intercellular spaces when tissue is frozen and thawed. *Opposite*: **bound water.**

free-living conditions Natural or greenhouse conditions experienced by **plantlets** upon transfer from ***in vitro*** conditions to soil. Prior to transfer, nutrients were supplied by the culture medium, but following transfer, plantlets must take up nutrients from soil and synthesize their own food supply.

freeze preservation *See*: **cryobiological preservation.**

freeze-dry The removal of water as vapour from frozen material under vacuum. Used to measure water content and to preserve samples, particularly spores. Unlike oven-drying, **bound water** remains associated with the specimen. *Synonym*: **lyophilize.**

fresh weight The weight, including the water content, of a specimen. *Synonym*: wet weight.

friable A term commonly used to describe a crumb-like callus. In this state, the callus is easily dissected and readily dispersed into single cells or clumps of cells in solution.

FSH *See*: **follicle stimulating hormone**.

functional food A foodstuff that provides a health benefit beyond basic nutrition, demonstrating specific health or medical benefits, including the prevention and treatment of disease.

functional gene cloning *See*: **candidate-gene strategy**.

functional genomics The field of research, that aims to determine patterns of **gene expression** and interaction in the genome, based on the knowledge of extensive or complete genomic **sequence** of an organism.

fungicide A chemical agent toxic to fungi.

fungus (pl.: fungi) Multinucleate single-celled or multicellular heterotrophic microorganisms, including yeasts, moulds, and mushrooms. They live as parasites, symbionts, or saprophytes. Lacking any vascular tissues (unlike plants), their cell walls are made of **chitin** or other non-**cellulose** compounds.

***Fusarium* spp.** A group of fungal pathogens of many economic crop species, particularly cereals, where severe infestation leads to losses in both grain yield and quality. The latter can be a particularly serious problem as many of these fungi produce mycotoxins, some of which are dangerous to both livestock and human health (*See:* **aflatoxin**). Specific strains are also employed on an industrial scale to produce **protein** for human consumption.

fusion biopharmaceuticals Fusion proteins with pharmaceutical properties. Their advantages are: 1. Synergistic activities in one molecule, i.e. when the molecule binds to its **target**, it can perform more than one function simultaneously; 2. An adverse effect or poor stability of one part of the molecule may be offset by the properties of the other; and 3. One part of the molecule can act as a targeting mechanism for the active **protein**. *See*: **immunotoxin, fusion toxin**.

fusion gene *See*: **chimeric gene**.

fusion protein A polypeptide translated from a **chimeric gene**. The different genes are joined so that their **coding** sequences are in the same **reading frame**, and the resulting **construct** is transcribed and translated as a single gene, producing a single **protein**. These are used for a number of purposes, including: 1. To add an **affinity tag** to a protein; 2. To produce a protein with the combined characteristics of two natural proteins; 3. To produce a protein where two different activities are physically linked. *See*: **fusion biopharmaceuticals.**

fusion toxin A fusion protein that consists of a toxic **protein domain** plus a cell **receptor** binding domain. The latter delivers the toxin directly to the **target** cell, thus sparing other healthy tissues from the effect of the **toxin**.

fusogenic agent Any chemical or virus, etc., that causes cells to fuse together.

G

G Abbreviation for **guanine**.

G cap The 5'-terminal methylated **guanine** nucleoside that is present on many eukaryotic **mRNAs**. It is joined to the m**RNA**, via a 5'?5' phosphodiester bond, after **transcription**. *See*: **cap site**.

G protein Proteins found on the inner surface of the **plasma** membrane, which bind to the **guanine** nucleotides, **GTP** and **GDP**. They transmit signals from outside the membrane, via trans-membrane (G-**protein**-coupled) receptors, to adenylate cyclase, which catalyses the formation of the second messenger, cyclic AMP, inside the cell.

galactomannan A gum in which the structural chain is made up of D-**mannose** units with 1?4 linkages. The ratio of galactose to mannose is 1:2.

gall A tumorous growth in plants. *See*: **crown gall**.

gamete A mature reproductive cell which is capable of fusing with a cell of similar origin but of opposite sex to form a **zygote** from which a new organism can develop. Gametes normally have a **haploid** chromosome content. In animals, a gamete is a **sperm** or **egg**; in plants, it is **pollen**, spermatic nucleus, or **ovum**.

gamete and embryo storage Storage of ova, **sperm** or fertilized embryos outside their original source. Almost invariably this means **cryopreservation**.

gametic (phase) disequilibrium In relation to any two loci, the occurrence of **haplotypes** (gametes) at a frequency other than that predicted from the product of the respective allele frequencies. *Opposite*: **gametic (phase) equilibrium**.

gametic (phase) equilibrium In relation to any two loci, the occurrence of **haplotypes** (gametes) at a frequency equal to the product of the frequency of the two relevant **alleles**. For example, *A* and *B* are in gametic equilibrium if the frequency of A_iB_i gametes equals the product of the frequencies of alleles A_i and B_i. *Opposite*: **gametic (phase) disequilibrium**.

gametoclone A plant regenerated from a **tissue culture** originating from gametic tissue.

gametogenesis The process of the formation of gametes.

gametophyte The phase of the plant **life cycle** that carries the **gamete** producing organs. In flowering plants, the **pollen grain** is the male gametophyte and the **embryo sac** is the female gametophyte.

gametophytic incompatibility A phenomenon in plants, in which a **pollen grain** is genetically incapable of fertilizing a particular **egg**, because both gametes carry the identical allele at an **incompatibility** locus (usually denoted *S*). This is a mechanism for forcing crossfertilization.

gap A missing section on one of the strands of double-stranded **DNA**. The **DNA** will therefore have a single-stranded region.

gapped DNA A **double-stranded DNA** molecule with one or more internal single-stranded regions.

gas transfer The rate at which gases are transferred from gas into solution, an important parameter in **fermentation** systems because it controls the rate at which the organism can metabolize. Efficient gas transfer can be achieved in several ways, including the use of small bubbles, from which gas dissolves faster than from larger ones, due to their larger surface area per unit of volume; or spreading the liquid out, for example, in a thin sheet, or in a thin **permeable** tube, as in **hollow fibre** bioreactor.

gastrula An early animal **embryo** consisting of two layers of cells; an embryological stage following the **blastula**.

GC island A segment of **double-stranded DNA** that is rich in GC base pairs. This type of sequence is characteristic of eukaryotic genomic regions with a high gene content.

GDP Abbreviation for **guanosine** 5'-diphosphate.

gel A jelly-like solid, used widely as a matrix for the **electrophoresis** of macromolecules, for **encapsulation**, and to solidify media for cell cultures.

gel electrophoresis *See*: **electrophoresis**.

gel filtration A method of **protein** or **DNA** purification, where differences in size are used to separate the components of a complex mixture.

gelatin A glutinous, **protein**aceous gelling and solidifying agent. Gelatin is produced by the partial **hydrolysis** (via boiling) of collagen, found in the connective tissues of many farm animals. Used to gel or solidify nutrient solutions for tissue culture, and as a food additive.

gelatinization The swelling of **starch** when added to hot water. **Hydrolysis** causes the molecule to lose structure, and technically gelatinization is not complete until there is no structure left at all.

Gelrite™ The brand name of a *Pseudomonas*-derived refined **polysaccharide** used as a gelling agent and **agar** substitute.

GEM Abbreviation for genetically engineered microorganism. *See*: **genetically modified**

organism.

gene (resources) conservation The conservation of species, populations, individuals or parts of individuals, by *in situ* or *ex situ* methods, to provide a diversity of genetic materials for present and future generations.

gene addition The addition of a functional copy of a gene to the **genome** of an organism.

gene amplification The selective production of multiple copies of one gene without a proportional increase in others.

gene bank 1. The physical location where collections of genetic material in the form of seeds, tissues or reproductive cells of plants or animals are stored. 2. Field gene bank: A facility established for the *ex situ* storage and maintenance, using horticultural techniques, of individual plants. Used for species whose seeds are **recalcitrant**, or for clonally propagated species of agricultural importance, e.g. apple varieties. 3. A collection of cloned **DNA** fragments from a single genome. Ideally the bank should contain cloned representatives of all the **DNA** sequences in the genome. 4. *See*: **library**.

gene cloning The synthesis of multiple copies of a chosen **DNA** sequence using a bacterial cell or another organism as a host. The gene of interest is inserted into a **vector**, and the resulting **recombinant DNA** molecule is amplified in an appropriate host cell. *Synonym*: **DNA cloning**.

gene construct *See*: **construct**.

gene conversion A process, often associated with recombination, during which one allele is replicated at the expense of another, leading to non-Mendelian **segregation** ratios.

gene expression The process by which a gene produces **mRNA** and **protein**, and hence exerts its effect on the **phenotype** of an organism.

gene flow The spread of genes from one breeding **population** to another (usually) related population by migration, thereby generating changes in **allele frequency**.

gene frequency *See*: **allele frequency**.

gene gun *See*: **biolistics**.
gene imprinting The differential expression of a single gene according to its parental origin.
gene insertion The incorporation of one or more copies of a gene into a chromosome.
gene interaction The modification of the action of one gene by another, non-allelic gene.
gene knockout *See*: **knockout**.
gene library *See*: **library**.
gene linkage *See*: **linkage**.
gene machine See: **transposon tagging**.
gene mapping *See*: **mapping**.
gene modification Chemical change to a gene's **DNA** sequence.
gene pool 1. The sum of all genetic information in a breeding **population** at a given time. 2. In **plant genetic resources**, use is made of the terms 'primary', 'secondary' and 'tertiary' gene pools. In general, members of the primary gene pool are inter-fertile; those of the secondary can be crossed with those in the primary gene pool under special circumstances; but to introgress variation from the tertiary gene pool, special techniques are required to achieve crossing.
gene probe *See*: **probe**.
gene recombination *See*: **recombination**.
gene regulation The process of controlling the synthesis or suppression of gene products in specific cells or tissues.
gene replacement The incorporation of a **transgene** into a **chromosome** at its normal location by **homologous** recombination, thus replacing the copy of the gene originally present at the locus.
gene sequencing *See*: **DNA sequencing**.
gene shears *See*: **ribozyme**.
gene silencing *See*: **silencing**.
gene splicing *See*: **splicing** (1).
gene stacking *See*: **stacked genes**.
gene The unit of **heredity** transmitted from generation to generation during sexual or asexual reproduction. More generally, the term is used in relation to the transmission and **inheritance** of particular identifiable traits. The simplest gene consists of a segment of **nucleic**

acid that encodes an individual **protein** or **RNA**.

gene therapy The proposed treatment of an inherited disease by the **transformation** of an affected individual with a wild-type copy of the defective gene causing the disorder. In germ-line (or heritable) gene therapy, reproductive cells are transformed; in somatic-cell (or non-inheritable) gene therapy, cells other than reproductive ones are modified.

gene tracking Following the **inheritance** of a particular gene from generation to generation.

gene transfer *See*: **transformation.**

gene translocation The movement of a gene from one chromosomal location to another.

genera Plural form of **genus.**

generally regarded as safe (Abbreviation: GRAS). Designation given to foods, drugs, and other materials with a long-term history of not causing illness to humans, even though formal **toxicity** testing may not been conducted. Certain host organisms for **recombinant DNA** experimentation have recently been given this status.

generation time *See*: **cell generation time.**

generative nucleus In many flowering plants, shed **pollen** is two-celled (in others it is three-celled or has a variable number). Before pollen is shed, the male **gametophyte** divides mitotically to give a generative and a vegetative nucleus. The former is the progenitor of the **sperm** cells.

generative *See*: **germ line.**

genet The individual(s) descended vegetatively from a single sexually produced zygote, including all entities derived from it. All these individuals are genetically identical to one another (barring mutation).

genetic assimilation Eventual extinction of a natural **species** as massive **gene flow** occurs from a related species.

genetic code The correspondence between the set of 64 possible **nucleotide triplets** and the **amino acids** and **stop codons** that they specify. *See* annex 3.

genetic complementation When two **DNA**

molecules that are in the same cell together produce a function that neither **DNA** molecule can supply on its own.

genetic disease A disease caused by an abnormality in the genetic material, which could be at the level of **DNA sequence** at a locus, or at the level of **karyotype**. Usually refers to inherited diseases, although somatic mutations can also cause disease without being inherited.

genetic distance A measure of the genetic similarity between any pair of populations. This is measured on the basis of **variation** in a combination of phenotypic traits, allele frequencies or **DNA** sequences. For example, the genetic distance between two populations having the same allele frequencies at a particular locus, and based solely on that locus, is zero.

genetic distancing The collection of the data on phenotypic traits, **marker** allele frequencies or **DNA** sequences for two or more populations, and estimation of the genetic distances between each pair of populations.

genetic diversity The heritable variation within and among populations which is created, enhanced or maintained by evolutionary or selective forces.

genetic drift Change in **allele frequency** from one generation to another within a population, due to the sampling of finite numbers of genes that is inevitable in all finite-sized populations. The smaller the population, the greater is the genetic drift, with the result that some alleles are lost, and **genetic diversity** is reduced. Thus minimization of genetic drift is an important consideration for **conservation** of genetic resources.

genetic engineering Modifying genotype, and hence phenotype, by **transgenesis**.

genetic equilibrium The maintenance of a **steady state** with respect to allele frequencies in a group of interbreeding organisms.

genetic erosion The loss over time of **allelic** diversity, particularly in farmed organisms, caused by either natural or man-made processes. *See*: **genetic drift**.

genetic fingerprinting *See*: **DNA fingerprinting.**

genetic gain The increase in productivity achieved following a change in **gene frequency** effected by selection.

genetic heterogeneity Occurs where the genetic determination of a given **phenotype** differs between individuals.

genetic immunization Delivery to a host organism of a cloned **gene** that encodes an **antigen**. After the cloned gene is expressed, it elicits an **antibody** response that protects the organism from infection by the relevant **pathogen.**

genetic information Information contained in a **nucleotide** base sequence in chromosomal **DNA** or **RNA.**

genetic linkage *See*: **linkage.**

genetic map The linear array of genes on a chromosome, based on **recombination** frequencies (linkage map) or physical location (physical or chromosomal map). *See*: **linkage map**.

genetic mapping *See*: **mapping.**

genetic marker A **DNA** sequence used to identify a particular location (locus) on a particular chromosome. *See*: **marker gene.**

genetic pollution Uncontrolled spread of **genetic information** (frequently referring to **transgenes**) into the genomes of organisms in which such genes are not present in nature.

genetic polymorphism *See*: **polymorphism.**

genetic relatedness A quantitative estimate of the proportion of genes, →, shared between the genomes of any two individuals, groups or populations, e.g. r = 0.5 for full siblings and parent **offspring** pairs.

genetic resources genetic material of actual or potential value.

genetic selection The process of selecting genes, cells, clones, etc., within populations or between populations or species. Genetic selection usually results in differential survival rates of the various genotypes, reflecting many variables, including the selection pressure and genetic variability present in populations.

genetic transformation *See*: **transformation.**

genetic use restriction technology (Abbreviation: GURT). A proposed technology applying **transgenesis** to genetically compromise the fertility or the performance of saved seed of a **cultivar** or of second generation animals. The intention is to protect the market for the seed producer or to prevent undesired escape of genes. Two types of GURTs have been patented: variety-level GURT (V-GURT), which produces **sterile** progeny, and **trait-**specific GURT (T-GURT), in which only the added value transgenic trait is genetically protected. *See*: **terminator gene, disrupter gene.**

genetic variation Differences between individuals attributable to differences in genotype.

genetically engineered organism (Abbreviation: GEO). Occasional alternative term for **genetically modified organism.**

genetically modified organism (Abbreviation: GMO). An organism that has been transformed by the insertion of one or more **transgenes.**

genetics The science of heredity.

genome 1. The entire complement of genetic material (genes plus non-coding sequences) present in each cell of an organism, **virus** or organelle. 2. The complete set of chromosomes (hence of genes) inherited as a unit from one parent.

genomic library A clone **library** specifically constructed from restriction fragments of the genomic **DNA** of an organism.

genomics The research strategy that uses molecular characterization and cloning of whole genomes to understand the structure, function and **evolution** of genes and to answer fundamental biological questions. *See*: **bioinformatics, functional genomics** and **proteomics.**

genotype 1. The genetic constitution of an organism. 2. The **allelic** constitution at a particular locus, e.g. *Aa* or *aa*. 3. The sum effect of all loci that contribute to the expression of a trait.

genus (pl.: genera) A group of closely related **species**, whose perceived relationship is typically based on physical resemblance, now often supplemented with **DNA sequence** data.

GEO Abbreviation for **genetically engineered organism**. *See*: **genetically modified organism.**

geotropism A growth curvature induced by gravity. *Synonym*: gravitropism.

germ 1.The botanical term for a plant embryo. 2. Colloquial: a disease-causing microorganism.

germ cell A member of a cell **lineage** (the **germ line**) leading to the production of gametes. In mammals, germ cells are found in the germinal epithelium of the ovaries and testes. *Synonym*: **germ line cell**. *Opposite*: **somatic cell.**

germ cell gene therapy The repair or replacement of a defective **gene** within the gamete-forming tissues, resulting in a heritable change in an organism's genetic constitution.

germ layer The layers of cells in an animal **embryo** at the **gastrula** stage, from which the various organs of the animal's body will be derived.

germ line A lineage of cells which, during the **development** of an organism, are set aside as potential gamete-forming tissues. The location, nature and time of formation of potential gamete-forming tissues are species specific, and may vary greatly from one species to another. *See:* **somatic**

germ line cell *See*: **germ cell.**

germ line gene therapy The delivery of a gene or genes to a fertilized **egg** or an early embryonic cell. The transferred gene(s) is present in all or some of the nuclei of the cells of the mature individual, including possibly the reproductive cells, and alters the **phenotype** of the individual that develops.

germicide Any chemical agent used to control or kill any pathogenic and non-pathogenic microorganisms.

germinal epithelium 1. A layer of epithelial cells on the surface of the **ovary** that are continuous with the mesothelium. 2. The layer of epithelial cells lining the seminiferous tubules of the testis, which gives rise to spermatogonia. *See*: **spermatogenesis**.

germination 1. The initial stages in the growth of a **seed** to form a seedling. 2. The growth of spores (fungal or algal) and **pollen** grains.

germplasm 1. An individual, group of individuals or a clone representing a genotype, variety, **species** or culture, held in an *in situ* or *ex situ* collection. 2. Original meaning, now no longer in use: the genetic material that forms the physical basis of **inheritance** and which is transmitted from one generation to the next by means of the germ cells.

gestation The period between conception (fertilization of the **egg**) to **parturition** (birth) spent *in utero* by the **foetus** of **viviparous** animals.

GFP Abbreviation for **green fluorescent protein**.

GH Abbreviation for **growth hormone**.

gibberellins A class of **plant growth regulators** which are active in the elongation, enhancement of flower, fruit and leaf size, germination, **vernalization** and other physiological processes.

gland A specialized group of cells or a single **cell** in animals or plants that secretes a specific substance. The two types of animal glands are: endocrine, which secrete directly into the blood vessels; and exocrine, which secrete through a duct or network of ducts into a body cavity or onto the body surface.

glaucous A surface with a waxy, white coating. In most cases, this waxy covering can be rubbed off.

globulins Common class of **proteins** in blood, eggs and milk, and seeds. Characterized by their slight solubility in water but are freely soluble in dilute salt solutions. Gamma-globulins are defined further by their electrophoretic behaviour, and include the **immunoglobulins**.

GLP Abbreviation for **good laboratory practice**.

glucocorticoid A steroid hormone that regulates **gene expression** in higher animals.

glucose invertase An enzyme that catalyses the **hydrolysis** of sucrose into its component monosaccharides, glucose and fructose.

glucose isomerase An enzyme that catalyses the interconversion of the two sugars, glucose and fructose. As fructose is a lower energy compound compared with glucose, a mixture of glucose and fructose with the enzyme will end up almost entirely as fructose.

glucosinolates A class of molecules produced in the seeds and green **tissue** of a range of plants, in particular brassicas. Their natural role is thought to be involved in plant-insect interactions. Their importance in plant **breeding** is largely because of their negative influence on taste and their positive effect on the prevention of cancers of the alimentary tract.

glucuronidase *See*: **beta-glucuronidase**.

gluten A mixture of two seed storage **protein** classes, gliadin and glutenin, found in the **endosperm** of cereal (particularly wheat) grain. High levels of gluten impart elasticity to dough, and thus the composition of wheat glutens largely determines whether a specific flour is suitable for biscuit or bread making. Sensitivity of the lining of the intestine to gluten in some humans results in coeliac disease, a condition that requires a gluten-free diet.

glycoalkaloids A group of modified alkaloids, including solanine and tomatine, having a range of toxic effects in humans and other species. They are of particular significance in food plants from the *Solanaceae*.

glycoform One of several structures possible for a given **glycoprotein**, determined by the type and position of attachment of the component **oligosaccharide**(s). Certain glycoforms may exhibit different biological activities from one another because the oligosaccharide units mediate interactions with other cell components.

glycolysis The sequence of reactions that converts glucose into pyruvate, with the concomitant production of **ATP**.

glycoprotein A **protein** molecule modified by the addition of one or several **oligosaccha-**

ride groups.

glycoprotein remodelling The use of restriction endoglycosidases to enzymatically remove **oligosaccharide** branches from glyco**protein** molecules. Removal of one or more oligosaccharide branches can lessen or abolish the antigenicity of the glyco**protein**, so allowing its injection for pharmaceutical purposes without incurring an unwanted immune response. *See*: **glycoform**.

glycosylation The covalent addition of sugar or sugar-related molecules to other classes of molecule, including **proteins** or nucleic acids.

glyphosate An active ingredient in some herbicides, killing plants by inhibiting the activity of plant **enolpyruvyl-shikimate 3-phosphate synthase**.

glyphosate oxidase An enzyme which catalyses the break-down of **glyphosate**, discovered in a **strain** of *Pseudomonas* bacteria which were found to produce unusually large amounts of the enzyme. The gene responsible has been incorporated into a variety of crop plants to enable them to tolerate applications of glyphosate-containing herbicides. It has also been used in conjunction with the **CP4 EPSPS** gene.

glyphosate oxidoreductase An enzyme from the **microorganism** *Ochrobactrum anthropi*, which catalyses the break-down of **glyphosate**. If the encoding **gene** (called *goxv247*) is inserted and properly expressed in a plant, these plants become tolerant of the application of glyphosate- and/or sulfosate-containing herbicides. An alternative to **CP4 EPSPS** or **glyphosate oxidase** encoded glyphosate tolerance.

GM food Abbreviation for genetically modified food. Food that contains above a certain legal minimum content of raw material obtained from **genetically modified organisms**.

GMO Abbreviation for **genetically modified organism**.

GMP Abbreviation for 1. **guanosine** 5'-monophosphate. *Synonym*: **guanylic acid**. 2. **good manufacturing practice**.

gobar *See*: **biogas.**

golden rice A biotechnology-derived rice, which contains large amounts of beta **carotene** (a precursor of **vitamin** A) in its seeds. Achieved by inserting two genes from daffodil and one from the bacterium *Erwinia uredovora*.

Golgi apparatus An assembly of vesicles and folded membranes within the **cytoplasm** of plant and animal cells that stores and transports secretory products (such as enzymes and hormones) and plays a role in formation of a **cell wall** (when this is present).

gonad One of the (usually paired) animal organs that produce reproductive cells (gametes). The most important gonads are the male testis, which produces spermatozoa, and the female ovary, which produces ova (egg cells). The gonads also produce hormones that control secondary sexual characteristics.

good laboratory practice (Abbreviation: GLP). Written codes of practice designed to reduce to a minimum the chance of procedural or instrument problems which could adversely affect a research project or other laboratory work.

good manufacturing practice (Abbreviation: GMP). Codes of practice designed to reduce to a minimum the chance of procedural or instrument/manufacturing plant problems which could adversely affect a manufactured product.

G-protein coupled receptor *See*: **G protein.**

graft 1. Verb. To place a detached branch or **bud (scion)** in close cambial contact with a rooted **stem (rootstock)** in such a manner that scion and rootstock unite to form a single plant. 2. Noun. Colloquial synonym for **scion**. *See*: **grafting, graft chimera, graft hybrid.**

graft chimera A plant which is a **mosaic** of two sorts of tissue differing in genetic constitution and assumed to have arisen as the result of a nuclear fusion following **grafting**. *See*: **graft hybrid.**

graft hybrid An individual formed from **graft** (2) and **stock** showing the characteristics of both progenitors. *See*: **graft chimera.**

graft inoculation test A test based on the use of a suspected viral carrier which is grafted to an indicator plant. If symptoms appear in the indicator plant, the viral **assay** is positive.

graft union The point at which a **scion** from one plant is joined to a **rootstock** from another plant.

grafting The process of making a **graft** (1).

graft-versus-host disease The rejection of transplanted organs by the recipient's immune system, due to attack of the recipient's T **lymphocytes** on the transplanted organ caused by differences in **major histocompatibility complex** proteins.

Gram staining A technique to distinguish between two major bacterial groups, based on whether or not their cell wall retains the Gram stain. Gram-positive bacteria are stained dark purple, while Gram-negative bacteria are only faintly coloured. Stain retention is determined by the structure of the **cell wall**.

granum (pl.: grana) Structure within the **chloroplasts**, appear as green granules with the light microscope and as a series of parallel lamellae with the electron microscope. They contain the **chlorophyll** and **carotenoid** pigments directly involved in photosynthesis.

GRAS Abbreviation for **generally regarded as safe**.

gratuitous inducer A substance that can induce **transcription** of a **gene** or genes, but is not a **substrate** for the induced enzyme(s).

gravitropism *See*: **geotropism**.

green fluorescent protein (Abbreviation: GFP). A **protein** derived from a species of jelly fish, that fluoresces when exposed to ultra violet light. Its encoding gene has been isolated and is replacing **GUS** as a **reporter gene** in plant transgenesis, since it can be assayed non-destructively in real time.

green revolution Name given to the dramatic increase in crop **productivity** during the third quarter of the 20th century, as a result of integrated advances in **genetics** and plant breeding, agronomy, and pest and disease control.

Gro-luxä A wide-spectrum fluorescent lamp suitable for artificial light for plant growth.

growth cabinet An enclosed space in which environmental conditions can be controlled. The degree of control over temperature, light and humidity is a function of the quality of the cabinet.

growth curve *See*: **growth phase**.

growth factor Any of various chemicals, particularly **polypeptides**, that have a variety of important roles in the stimulation of new **cell** growth and cell maintenance. They bind to the cell surface on receptors. Specific growth factors can cause new cell proliferation.

growth hormone (Abbreviation: GH). A group of hormones, secreted by the mammalian pituitary gland, that stimulates **protein synthesis** and growth of the long bones in the legs and arms. They also promote the breakdown and use of fats as an energy source, rather than glucose. *Synonym*: **somatotropin**.

growth inhibitor Any substance inhibiting the growth of an organism. The inhibitory effect can range from mild inhibition (growth retardation) to severe inhibition or death (toxic reaction). The concentration of the inhibitor, the length of exposure to it, and the relative susceptibility of the organisms exposed to the inhibitor, are all important factors which determine the extent of the inhibitory effect.

growth phase Each of the characteristic periods in the **growth curve** of a bacterial culture, as indicated by the shape of a graph of viable **cell number** versus time, namely: **lag phase**; **logarithmic phase**; **stationary phase**; **death phase**.

growth rate Change in an organism's mass per unit of time.

growth regulator A synthetic or natural compound that at low concentrations elicits and controls growth responses in a manner similar to hormones.

growth retardant A chemical that selectively interferes with normal hormonal promotion of growth and other physiological processes, but without appreciable toxic effects.

growth ring Rings visible in a cross-section of a woody **stem**, such as a tree trunk. Each ring represents the **xylem** formed in one year as a result of fluctuating activity of the **vascular** cambium.

growth substance Any organic substance, other than a nutrient, that is synthesized by plants and regulates growth and development. They are usually made in a particular region, such as the shoot tip, and transported to other regions, where they take effect.

GTP Abbreviation for **guanosine** 5'-triphosphate, a **nucleotide** which is important as a **ligand** for **G proteins** and as a direct precursor molecule for **RNA** synthesis. *See*: **guanylic acid**.

guanine (Abbreviation: G). One of the **bases** found in **DNA** and **RNA**. *See*: **guanosine**.

guanosine The (ribo)**nucleoside** resulting from the combination of the **base** guanine (G) and a D-**ribose** sugar. The corresponding **deoxyribonucleoside** is called deoxyguanosine. *See*: **GTP**, **dGTP**, **guanylic acid**.

guanosine triphosphate (guanosine 5-triphosphate) Abbreviation: **GTP**. *See*: **guanylic acid**.

guanylic acid Synonym for **guanosine** monophosphate (abbreviation: GMP), a (ribo)**nucleotide** containing the **nucleoside guanosine**. The corresponding **deoxyribonucleotide** is called deoxyguanylic acid.

guard cell Specialized epidermal cells found in pairs around a **stoma**. Their function is to control the opening and closing of the stoma through changes in turgor.

guide RNA An RNA molecule that contain sequences that function as a **template** during **RNA** editing. *See*: **guide sequence**.

guide sequence An **RNA** molecule (or a part of it) which hybridizes with eukaryotic **mRNA** and aids in the **splicing** of **intron** sequences. Guide sequences may be either external (EGS) or internal (IGS) to the RNA being processed and may hybridize with either intron or **exon** sequences close to the splice junction. *See*: **split gene**.

GURT Abbreviation for **genetic use restriction technology**.

GUS Abbreviation for **beta-glucuronidase**.

gus gene An *E. coli* gene that encodes for production of **beta-glucuronidase** (GUS). Because this activity is absent in plants, the gene is commonly utilized as a **reporter gene** to detect the occurrence of **transformation** events.

gymnosperm A class of plant (e.g. conifers) whose ovules and the seeds into which they develop are borne unprotected, rather than enclosed in ovaries, as are those of the flowering plants, the (**angiosperms**).

gynandromorph An individual in which one part of the body is female and another part is male; a sex mosaic.

gynogenesis Female **parthenogenesis**: after fertilization of the ovum, the male **nucleus** is eliminated and the **haploid** (gynogenetic) individual possesses the maternal genome only.

gyrase *See*: **DNA helicase**.

H

h Prefix used to designate the human form of an enzyme. For example, **hGH** is human growth hormone.

habituation The phenomenon whereby, after a number of sub-cultures, cells can grow without the addition to the **tissue culture** medium of previously obligatory factors. Such cells are then **autonomous**.

HAC Abbreviation for **human artificial chromosome**.

haemoglobin **Protein** containing iron, located in erythrocytes of vertebrates; important for the transportation of oxygen to the cells of the body.

haemolymph The mixture of blood and other fluids in the body cavity of an invertebrate.

haemophilia A sex-linked hereditary bleeding disorder in which it takes a long time for the blood to clot and abnormal **bleeding** occurs. This disease affects mostly males.

hairpin loop A region in one strand of a **polynucleotide** which, due to an **inverted repeat** in the sequence, may under appropriate conditions fold back on itself and form a limited segment of **double-stranded DNA** with a loop at one end.

hairy root culture A culture consisting of highly branched roots of a plant. A plant tissue is treated with the bacterium ***Agrobacterium rhizogenes*** containing the **Ri plasmid**, which causes the explant to grow highly branched roots from the sites of infection. **Transgenes** engineered into the plasmid can be expressed in these cultures.

hairy root disease A disease of broad-leaved plants, where a **proliferation** of root-like tissue is formed from the stem. Hairy root disease is a tumorous state similar to **crown gall**, and is induced by the bacterium ***Agrobacterium rhizogenes,*** when containing an **Ri plasmid**.

halophyte A plant **species** adapted to soils containing a concentration of salt that is toxic to most plant species. *See*: salt tolerance.

hanging droplet technique *See*: microdroplet array.

haploid A cell or organism containing one of each of the pairs of **homologous** chromosomes found in the normal **diploid** cell.

haplotype A specific **allelic** constitution at a number of loci within a defined **linkage** block.

haplozygous *See*: **hemizygous.**

hapten A small molecule, which by itself is not an **antigen**, but which as a part of a larger structure when linked to a carrier **protein**, can serve as an **antigenic determinant**.

haptoglobin A serum **protein** that interacts with **haemoglobin** during recycling of the iron molecule of haemoglobin. *Synonym*: alpha globulin.

hardening off Adapting glasshouse or **controlled environment** grown plants to outdoor conditions by reducing availability of water, lowering the temperature, increasing light intensity, or reducing the nutrient supply. The hardening-off process conditions plants for survival when transplanted outdoors.

Hardy-Weinberg equilibrium The frequencies of **genotypes** at a locus resulting from random mating at that locus; for two **alleles**, A_1 and A_2, with respective frequencies in a **population** of p and q, the Hardy-Weinberg equilibrium frequencies are p^2 A_1A_1; $2pq$ A_1A_2; q^2 A_2A_2. Departure from these frequencies is an indication that random mating is not occurring.

harvesting 1. The process involved in gathering ripened crops. 2. The collection of cells from cell cultures or of organs from donors for the purpose of transplantation.

heat shock protein (Abbreviation: HSP). A class of **protein chaperones** which are typically over-expressed as a response to heat **stress**. Two such **proteins** - HSP 90 and HSP 70 - have a role in ensuring that crucial **proteins** are folded into the correct conformation. *Synonym*: stress **protein**.

heat therapy *See*: **thermotherapy**.

helix A structure with a spiral shape. The normal state **of double-stranded DNA** is in the form of a **double helix.**

helminth A class of parasitic worms, especially those which are internal parasites of man and animals.

helper cell T cells that assist in stimulating B and T **lymphocytes** to develop into antibody-producing **plasma cells** and **killer T cells,** respectively.

helper plasmid A **plasmid** that provides a function or functions to another plasmid in the same cell.

helper T cell *See*: **helper cell.**

helper T lymphocyte *See*: **helper cell.**

helper virus A virus that provides a function or functions to another virus in the same cell.

hemicellulase An enzyme that catalyses the degradation of **hemicellulose.**

hemicellulose Any cellulose-like carbohydrate, but excluding **cellulose** itself. Together with **pectin** and lignin, hemicelluloses form the **cell wall** matrix.

hemizygous The condition in which genes are present only once in the genotype and not in pairs. Occurs for all genes in **haploids**, for all genes located in the differential segments of the sex chromosomes in diploids, and in various **aneuploids** and **deletion** mutant **heterozygotes**. *Synonym*: **haplozygous**.

hemoglobin *See*: **haemoglobin**.

hemolymph *See*: **haemolymph**.

hemophilia *See*: **haemophilia**.

HEPA filter Abbreviation for **high efficiency particulate air filter**. A filter capable of excluding particles larger than 0.3ìm. HEPA filters are used in laminar air flow cabinets to ensure that the air is pathogen-free. *See*: **prefilter**.

herbicide A substance that is toxic to plants; the active ingredient in agrochemicals intended to kill specific unwanted plants, especially weeds.

herbicide resistance The ability of a plant to remain unaffected by the application of a herbicide.

heredity Resemblance among individuals related by descent; transmission of traits from parents to offspring.

heritability The degree to which a given **trait** is controlled by inheritance, as opposed to being controlled by non-genetic factors. *See*: **broad-sense heritability; narrow-sense heritability**.

hermaphrodite 1. An animal that has both male and female reproductive organs, or a mixture of male and female attributes. 2. A plant whose flowers contain both **stamen** and carpels. *Synonym*: intersex.

heteroallele A gene having mutations at two or more different sites.

heterochromatin Regions of chromosomes that remain contracted during **interphase** and therefore stain more intensely in cytological preparations. These regions have a high content of **repetitive DNA**, and a low content of genes; thus they are for the most part genetically inactive. *Opposite*: **euchromatin**.

heteroduplex A double-stranded **DNA** molecule or **DNA-RNA** hybrid, where each strand is of a different origin. Where the two DNAs differ significantly in sequence, single-stranded regions will be revealed when the heteroduplex is observed under the electron microscope. A **map** of **homologous** and non-homologous regions of the two molecules may thereby be constructed (heteroduplex mapping). *Synonym*: **hybrid DNA** (DNA/RNA). *See*: **heteroduplex analysis**.

heteroduplex analysis The use of the electrophoretic mobility of **heteroduplex DNA** to estimate the degree of non-homology between the sequences of the two strands. Gel mobility tends to be reduced as the amount of **sequence** divergence increases, because the effective size of a fully complementary pair of strands is smaller than that of a partially complementary structure.

heterogametic Producing unlike gametes with regard to the sex chromosomes. In mammals, the XY male is heterogametic, and the XX female is **homogametic**.

heterogeneity *See*: **genetic heterogeneity**.

heterogeneous nuclear RNA (Abbreviation: hn**RNA**). Large **RNA** molecules, which are found in the **nucleus** of a eukaryotic **cell** and the precursors of **mRNA** and other **RNA** molecules.

heterokaryon A cell with two or more different nuclei as a result of **cell** fusion. *See*: **synkaryon**.

heterologous From a different source.

heterologous probe A **DNA probe** that is derived from one species and used to screen for a similar **DNA** sequence from another species.

heterologous protein *See*: **recombinant protein**.

heteroplasmy A condition in which two genetically different organelles are present in the same cell. The equivalent to **heterozygous** in the context of nuclear genes. *Opposite*: **homoplasmy**.

heteroploid Cells with nuclei containing **chromosome** numbers other than **diploid**.

heteropyknosis The property of certain chromosomes, or of their parts, to remain more dense during the **cell cycle** and to stain more intensely than other chromosomes or parts.

heterosis *See*: **hybrid vigour**.

heterotroph Organism non capable of self-nourishment utilizing carbon dioxide or carbonates as the sole source of carbon and obtaining energy from radiant energy or from the oxidation of inorganic elements, or compounds such as iron, sulphur, hydrogen, ammonium and nitrites. *Opposite*: **autotroph**.

heterotrophic (adj.) *See:* **heterotroph**.

heterozygote An individual with non-identical alleles for a particular **gene** or genes. The condition is termed "heterozygous". *Opposite*: **homozygote.**

heterozygous (adj.) See: **heterozygote**.

Hfr High-frequency **recombination** strain of *Escherichia coli;* in these strains, the **F factor** (plasmid) is integrated into the bacterial chromosome.

hGH Abbreviation for human **growth hormone**.

high efficiency particulate air filter *See*: **HEPA filter**.

high throughput screening Automated systems designed to process large numbers of assays, especially in the context of genotyping.

histocompatibility complex *See*: **major histocompatibility complex.**

histocompatibility The degree to which **tissue** from one organism is tolerated by the immune system of another organism.

histoglobulin The peptides present on the surface of nucleate cells, responsible for the differences between genetically non-identical individuals that cause rejection of **tissue grafts** between such individuals. Products of the **major histocompatibility complex** genes.

histology Science that deals with the microscopic structure of animal and plant tissues.

histone Group of water-soluble **proteins** rich in basic amino acids, closely associated with **DNA** in plant and animal **chromatin**. Histones are involved in the coiling of **DNA** in chromosomes and in the regulation of gene activity.

HLA Abbreviation for **human-leukocyte-antigen system**. *See*: **major histocompatibility antigens**.

hnRNA Abbreviation for **heterogeneous**

nuclear RNA.

Hogness box Synonym for **TATA box.**

hollow fibre A tube of porous material, with an internal diameter of a fraction of a millimetre, making its ratio of surface area to internal volume very large. Employed as filters or in bioreactors as a method of retaining cells while allowing the easy removal of **spent medium** and/or products.

holoenzyme *See*: **apoenzyme.**

holometabolous An insect that undergoes complete metamorphosis to the adult from a morphologically distinct larval stage.

homeobox A highly conserved 180 bp **DNA** sequence that controls body part-, organ- or tissue-specific **gene** expression, most particularly involved in segmentation in animals (e.g. development of antennae or legs of ***Drosophila melanogaster***), but also in a variety of other **eukaryotes**. It encodes a DNA-binding region, the **homeodomain**, which acts as a **transcription factor.**

homeodomain *See*: **homeobox.**

homeotic genes Genes that act in concert to determine fundamental patterns of development.

homeotic mutation A mutation that causes a body part to develop in an inappropriate position in an organism, such as the mutation in ***Drosophila melanogaster*** that causes legs to develop on the head in place of antennae.

homoallele One of a number of otherwise identical **alleles** which differ at the same site in their sequence. **Homoalleles** are inherited as strict alternatives; but **heteroalleles**, in principle, could through **recombination** create a **genotype** which contains a 'double' variant.

homodimer A **protein** comprising two identical **polypeptide** chains, or a **dimer** of identical residues.

homoduplex DNA A double-stranded fully complementary **DNA** molecule.

homoeologous Referring to **chromosomes** which are descended from a common progenitor, but which have evolved to be no longer fully **homologous**. Homoeologous chromosomes have similar **gene** content to one another, but are structurally altered in subtle ways to inhibit, and sometimes completely prevent their **pairing** with one another at meiosis.

homogametic Producing similar **gametes** with regard to the **sex chromosomes**. In mammals, the female is homogametic (XX), and the male is **heterogametic** (XY).

homogenotization An allele **replacement** technique, in which a bacterial cell is transformed with a plasmid containing an altered copy of the target sequence, and a double recombination event effects the substitution of the **wild type** allele by the altered one. An **antibiotic resistance** gene is usually fused to the altered copy in the **plasmid**, so that double recombinants can be selected.

homokaryon A cell with two or more identical nuclei as a result of fusion. *Opposite*: **heterokaryon.**

homologous 1. General definition: from the same source, or having the same evolutionary function or structure. 2. Of **chromosomes**: identical with respect to gene content and linear ordering. Homologous chromosomes pair and recombine with one another at **meiosis.** 3. Of **DNA/proteins**: identical, or nearly identical **nucleotide**/amino acid sequence.

homologous recombination The exchange of **DNA** fragments between the two non-sister **chromatids** of the same **chromosome** in the course of meiosis.

homology 1. The degree of identity between individuals, or characters. 2. The degree of identity of sequence (nucleotide or amino acid) between a number of **DNA** or **polypeptide** molecules.

homomultimer A **protein** consisting of a number of identical subunits.

homoozygous (adj.) See: **homozygote.**

homoplasmy The condition in which all copies of an **organelle** in a cell are genetically identical. *Opposite*: **heteroplasmy.**

homopolymer A polymer (nucleic acid, polypeptide, etc.) which contains only one kind of

residue (e.g. the **polynucleotide** GGGGGGGG...).

homopolymeric tailing *See*: **tailing.**

homozygote An individual that has two copies of the same **allele** for a given **gene** on its two **homologous chromosomes**. The condition is termed "homozygous". *Opposite*: **heterozygote.**

hormone A specific chemical, produced in one part of a plant or animal body, and transported to another part where, at low concentrations, it promotes, inhibits or quantitatively modifies a biological process.

host An organism that contains another organism or a **cloning vector.**

host-specific toxin A metabolite, produced by a **pathogen**, and which is responsible for the adverse effects of the pathogen. The toxin has a host **specificity** equivalent to that of the pathogen. Utilized for *in vitro* selection experiments to screen for **tolerance** or **resistance** to the pathogen.

hot spot *See*: **recombinational hot spot.**

HSA *See*: **serum albumin.**

HSP *See*: **heat shock protein.**

human artificial chromosome (Abbreviation: HAC). Analogous to **yeast artificial chromosome**, a construct comprising a human **centromere** and **telomeres**, which would allow for the cloning of very large fragments of **DNA**, and their transfer into human cells for the purpose of **gene therapy**. Not yet established as a working technology, although some partial success has been reported.

human growth hormone (Abbreviation: hGH). *See*: **growth hormone.**

human-leukocyte-antigen system (Abbreviation: HLA). *See*: **major histocompatibility antigens.**

humoral immune response *See*: **antibody-mediated immune response.**

Hup$^+$ Abbreviation for **hydrogen-uptake positive.**

hybrid 1. The offspring of two genetically unlike parents. 2. Of **DNA** molecules, *see*: **heteroduplex.**

hybrid arrested translation A method used to identify what **protein**(s) are encoded by a particular cloned **DNA** sequence. A total m**RNA** preparation, which contains many different m**RNA**s, is hybridized with cloned **DNA**, so that those **mRNA** molecules **homologous** to the cloned **DNA** will **anneal** to give **DNA/RNA heteroduplex**es. The non-annealed **mRNA** molecules can be translated *in vitro* and this is then compared to translation products from the untreated m**RNA**s. *See*: **hybrid released translation.**

hybrid cell *See*: **synkaryon.**

hybrid dysgenesis Infertility and an increased incidence of **chromosome** mutations thought to be caused by the activation of **transposons.**

hybrid released translation A method used to identify the gene product of a cloned gene. The cloned **DNA** is immobilized and hybridized with a mixed **mRNA** sample, so that only mRNA sequences **homologous** to the cloned DNA will be retained. These mRNA molecules are subsequently removed and translated *in vitro*. *See*: **hybrid arrested translation.**

hybrid seed 1. Seed produced by crossing genetically dissimilar parents. 2. In plant breeding, used colloquially for seed produced by specific crosses of selected pure lines, such that the $\mathbf{F_1}$ crop is genetically uniform and displays **hybrid vigour**. As the F_1 plants are **heterozygous** with respect to many genes, the crop does not breed true and so new seed must be purchased each season.

hybrid selection The process of choosing individuals possessing desired characteristics from among a hybrid population.

hybrid vigour The extent to which a hybrid individual outperforms both its parents with respect to one or many traits. The genetic basis of hybrid vigour is not well understood, but the phenomenon is widespread, particularly in **inbreeding** plant species. *Synonym*: **heterosis.**

hybridization 1. The process of forming a **hybrid** by **cross pollination** of plants or by mating animals of different types. 2. The production of **offspring** of genetically different parents, normally from sexual reproduction, but also asexually by the fusion of **protoplasts** or by **transformation**. 3. The **pairing** of two **DNA** strands, often from different sources, by hydrogen bonding between **complementary** nucleotides.

hybridoma A synthetic **hybrid** cell, derived by fusing a B **lymphocyte** with a tumour cell. The former secretes a single antibody, while the latter confers the property of growing indefinitely in tissue culture. The underlying technology behind the **monoclonal antibody**.

hydrogen-uptake positive (Abbreviation: Hup^+). A term describing a **microorganism** that is capable of assimilating (or taking up) hydrogen gas.

hydrolysis A chemical reaction in which water is added across a covalent bond, often cleaving the molecule into two. Occurs for example, when polynucleotides, polypeptides, and polysaccharides are broken into their component **monomers**. Thus, sucrose can be hydrolysed to glucose and fructose; and **proteins** to individual amino acids.

hydrophobic interaction An interaction between a hydrophobic ('water-hating') part of a molecule and an aqueous environment. Particularly significant in establishing the **conformation** of molecules in solution, and thus their biological activity. Many enzymes have a structure where the **polypeptide** chain is folded to form a hydrophobic core and a hydrophilic ('water-loving') surface.

hydroponics The growing of plants without soil. Plants are fed with an aerated solution of nutrients, and the roots are either supported within an inert matrix, or are freely floating in the nutrient solution.

hygromycin An **antibiotic** used as selective agent in bacterial and **transgenic** plant cell cultures.

hyperploid The situation in which a particular chromosome or **chromosome** segment is present in more than the normal number. *Opposite*: **hypoploid**.

hypersensitive response 1. A specific reaction of a plant to attack by a **pathogen**. The plant cells surrounding the point of **infection** rapidly die and dry out, so that pathogen spread within the plant is prevented. Often associated with the interaction of race-specific *R* **genes** with a matching pathogen avirulence. 2. The abnormal response of an animal to the presence of a particular **antigen.**

hypersensitive site Regions in the **DNA** that are highly **susceptible** to digestion with **restriction endonucleases**.

hypertonic A solution with an **osmotic potential** greater than that of living cells. Treatment with such solutions leads to the **plasmolysis** of cells. *Opposite*: **hypotonic**.

hypervariable region The parts of both the heavy and light chains of an **antibody** molecule that enable it to bind to a specific site on an **antigen**.

hypervariable segment A region of a **protein** that varies considerably between strains or individuals.

hypocotyl The portion of an **embryo** or seedling below the **cotyledons**. A transitional area between **stem** and root.

hypomorph A mutation that reduces, but does not completely abolish gene expression.

hypoplastic Defective and reduced growth or development (e.g. dwarfing and stunting in plants) resulting from an abnormal condition, for example, disease or nutritional stress.

hypoploid The situation in which a particular chromosome or **chromosome** segment is present in less than the normal number. *Opposite*: **hyperploid.**

hypothalamic peptides Peptides generated in the vertebrate forebrain and concerned with regulating the body's physiological state.

hypotonic Osmotic potential less than that of living cells. Cells placed in a hypotonic solu **I/E region** Abbreviation for **integration-excision region.**

I

ICSI *See:* **intracytoplasmic sperm injection.**

identical twin *See*: **monozygotic twin**.

idiogram *See*: **karyogram**.

idiotype An identifying property or characteristic of an item or system. 1.A plant form expected on physiological grounds to represent an optimal type for the environment in which the plant is to be grown. 2. A classification of **antibody** molecules according to the antigenicity of the variable regions. Each idiotype is unique to a particular **immunoglobulin** raised to a particular **antigen**.

IgA/Igd/IgG/IgE/IgM *See*: **antibody class**.

IGS Abbreviation for 1. **internal guide sequence**. 2. **intergenic spacer**.

imaginal disc A mass of cells in the larvae of *Drosophila melanogaster* and other **holometabolous** insects that gives rise to particular adult organs, such as antennae, eyes or wings.

imbibition 1. The absorption of liquids or vapours into the ultramicroscopic spaces or pores found in materials. 2. The initial water uptake by seeds prior to germination.

immediate early gene A viral gene that is expressed promptly after infection.

immobilized cells Cells entrapped in matrices such as alginate, polyacrylamide and agarose, for use in membrane and filter bioreactors.

immortalization The genetic **transformation** of a cell type into **a cell line** which can proliferate indefinitely.

immortalizing oncogene A gene that, upon **transfection**, enables a **primary cell** to grow indefinitely in culture.

immune response The processes, including the synthesis of **antibodies**, that are used by vertebrates to respond to the presence of a foreign **antigen**. *See*: **primary immune response**; **secondary immune response**.

immunity The lack of susceptibility of an animal or plant to **infection** by a particular **pathogen**, or to the harmful effects of their toxins.

immunization The production of **immunity** in an individual by artificial means. Active immunization involves the introduction, either orally or by infection, of specially treated bacteria, viruses or their toxins so as to stimulate the production of **antibodies**. *See*: **passive immunity**.

immunoaffinity chromatography A purification technique in which an **antibody** is bound to a matrix and is used to isolate a **protein** from a complex mixture. *See*: **affinity chromatography**.

immunoassay An detection system for a particular molecule, which exploits the specific binding of an **antibody** raised against it. For measurement, the **antibody** can incorporate a radioactive or fluorescent **label**, or be linked to an **enzyme** which catalyses an easily monitored reaction such as a change in colour (*see*: **ELISA**). *Synonym*: **immunodiagnostics**.

immunochemical control *See*: **immunotherapy**.

immunodiagnostics *See*: **immunoassay**.

immunogen *See*: **antigen**.

immunogenicity The ability to elicit an **immune response**.

immunoglobulin *See*: **antibody**.

immunoprophylaxis The process of active or passive **immunization**. Active immunization with **vaccines** leads to long-term protection through mobilizing the organism's own immune system. Passive immunization is achieved by injection of **antibody** against a specific pathogen obtained either by **fractionation** of blood from an individual previously infected with the **pathogen**, or by **monoclonal antibody** technology. *See*: **adoptive immunization, passive immunity**.

immunosensor A **biosensor** having an **antibody** as the biological element.

immunosuppression The suppression of the **immune response**. Necessary following organ transplants from a genetically different donor in order to prevent the host rejecting the grafted **organ**.

immunosuppressor A substance, an agent or a condition that prevents or diminishes the **immune response**.

immunotherapy The use of an **antibody** or a **fusion protein** containing the **antigen binding** site of an antibody to cure a disease or enhance the well-being of a patient. *Synonym*: **immunochemical control**.

immunotoxin Protein drugs consisting of an **antibody** joined to a **toxin** molecule. Made by either chemically linking the molecules, or by fusing the genes for the toxin and the antibody, and expressing a **fusion protein**. The antibody portion of the molecule targets the toxin.

impeller An agitator that is used for mixing the contents of a bioreactor.

in silico In a computer file. In the present context, the use of data bases of **DNA** and **protein** sequence to help answer biological questions. This is growing area of biology as the amount of **genomics** and **proteomics** data continues to grow. *See*: **bio-informatics**.

***in situ* colony hybridization** A procedure for screening bacterial colonies or **plaques** growing on plates or membranes for the presence of specific **DNA** sequences by the hybridization of **nucleic acid** probes to the **DNA** molecules present in these colonies or plaques. *Synonym*: ***in situ* plaque hybridization**.

***in situ* hybridization** The visualization of *in vivo* location of **macromolecules** (particularly polynucleotides and polypeptides) by the histological staining of tissue sections or cytological preparations via labelled probes/antibodies.

in situ In the natural place or in the original place. 1. Experimental treatments performed on cells or **tissue** rather than on extracts from them. 2. Assays or manipulations performed with intact tissues.

***in situ* plaque hybridization** *See*: ***in situ* colony hybridization**.

***in vitro* embryo production** (Abbreviation: IVEP). The combination of **ovum pickup**, ***in vitro* maturation** of ova, and ***in vitro* fertilization**. A potential means of overcoming the variability between donors in number of ova collected in embryo-transfer programmes.

***in vitro* fertilization** (Abbreviation: IVF). A widely used technique in human and animal science, whereby the **egg** is fertilized with **sperm** outside the body before re-implanting into the uterus.

***in vitro* maturation** (Abbreviation: IVM). Culture of immature ova in the laboratory, usually until they are ready for ***in vitro* fertilization**.

***in vitro* mutagenesis** *See*: **directed mutagenesis**.

in vitro Outside the organism, or in an artificial environment. Applied for example to cells, tissues or organs cultured in glass or plastic containers.

***in vitro* transcription** The cell-free synthesis from **DNA** or **RNA** in the test tube. *Synonym*: **cell-free transcription**.

***in vitro* translation** The synthesis of **proteins** from isolated **mRNA** molecules in the test-tube. *Synonym*: **cell-free translation**.

***in vivo* gene therapy** The delivery of a gene or genes to a **tissue** or **organ** of a complete living individual to alleviate a genetic disorder.

in vivo The natural conditions in which organisms reside. Refers to biological processes that take place within a living **organism** or **cell** under normal conditions.

inactivated agent A virus, bacterium or other organism that has been treated to prevent it from causing a disease. *See*: **attenuated vaccine**.

inbred line The product of **inbreeding**, i.e. the intercrossing of individuals that have ancestors in common. In plants and laboratory animals, it refers to populations resulting from at least 6 generations of selfing or 20 generations of brother-sister mating, so that they have become, for all practical purposes, completely **homozygous**. In farm animals, the term is sometimes used to describe populations that have resulted from several generations of the mating of close relatives, without having reached complete homozygosity.

inbreeding depression The reduction in vigour over generations of **inbreeding**. This affects species which are normally **outbreeding** and highly heterozygous. *See*: **hybrid vigour**.

inbreeding Matings between individuals that have one or more ancestors in common, the extreme condition being self-fertilization, which occurs naturally in many plants and some primitive animals. *Synonym*: **endogamy**.

inclusion body A **protein** that is overproduced in a **recombinant** bacterium and forms a crystalline structure inside the bacterial cell.

incompatibility 1. Genetically or physiologically determined prevention of intermating. 2. A physiological interaction resulting in **graft** rejection or failure. 3. A function of a related group of **plasmids**. Incompatible plasmids share similar **replication** functions, and this leads to the exclusion of one or the other plasmid if they are present in the same cell. Plasmids belonging to one **incompatibility group** are very closely related.

incompatibility group Plasmids must belong to different incompatibility groups to co-exist within the same cell. A **plasmid cloning vector** should always belong to an incompatibility group different from that of the host bacterium's **endogenous** plasmids.

incomplete digest *See*: **partial digest**.

incomplete dominance A gene action in which **heterozygotes** have a **phenotype** that is different from either **homozygote**, and is usually intermediate between them.

incomplete penetrance Where the **phenotype** does not allow perfect prediction of the **genotype** as a result of interference in **gene expression** by the environment.

incubation 1. The hatching of eggs by means of heat, either natural or artificial. 2. Period between **infection** and appearance of symptoms induced by a **pathogen**. 3. The **culture** of cells and organisms.

incubator An apparatus in which environmental conditions (light, photoperiod, temperature, humidity, etc.) are fully controlled; used for hatching eggs, multiplying microorganisms, culturing plants, etc.

indehiscent Describing a fruit or fruiting body that does not open to release its seeds or spores when ripe.

independent assortment The random distribution during **meiosis** of **alleles** (at different genes) to the gametes that is the case when the genes in question are located on different chromosomes or are unlinked on the same chromosome. *See*: **linkage**.

indeterminate growth The condition in which the **terminal bud** persists and produces successive lateral branches over an indefinite period. *Opposite*: **determinate growth**.

indirect embryogenesis Plant **embryo** formation on **callus** tissues derived from **explants**, including zygotic or somatic embryos and seedlings. *Opposite*: **direct embryogenesis**.

indirect organogenesis Plant organ formation on **callus** tissues derived from **explants**. *Opposite*: **direct organogenesis**.

inducer A low-molecular-weight compound or a physical agent that associates with a **repressor protein** to produce a complex that can no longer bind to the **operator**. Thus, the presence of the inducer turns on the expression of the gene(s) controlled by the operator.

inducible A gene or **gene** product whose **tran-**

scription or synthesis is increased by exposure of the cells to an **inducer** or to a condition, e.g. heat. *Opposite*: **constitutive**.

inducible enzyme An enzyme that is synthesized only in the presence of the **substrate** that acts as an inducer.

inducible gene A gene that is expressed only in the presence of a specific metabolite, the inducer.

inducible promoter The activation of a **promoter** in response to either the presence of a particular compound, i.e. the **inducer**, or to a defined external condition, e.g. elevated temperature.

induction media 1. Media used to induce the formation of organs or other structures. 2. Media causing variation or **mutation** in the tissues exposed to it.

induction The act or process of causing some specific effect to occur; for example, the **transcription** of a specific gene or operon, or the production of a **protein** by an organism after it is exposed to a specific stimulus.

inembryonation *See*: **artificial inembryonation**.

infection The successful colonization of any living organism by a **pathogen**.

infectious agent Synonym of **pathogen**.

infiltrate The entry of liquid into pores or other spaces.

inflorescence The flowers of a plant, and the way those flowers are arranged.

inheritance The transmission of genes and **phenotypes** from generation to generation.

inhibitor 1. Any substance or object that retards a chemical reaction. 2. A **metabolite** or modifier gene that interferes with a reaction or with the expression of another gene.

initial Cells in a **meristem** that remain permanently capable of differentiation, and which develop into tissues of particular structure and function.

initiation Causing something to start. 1. Early steps or stages of a **tissue culture** process. 2. Early stages of biosynthesis.

initiation codon *See*: **start codon**.

initiation factor Soluble **protein** required for the

initiation of **translation**.

inoculate Deliberately introduce, in contrast to contamination. 1. In bacteriology, **tissue culture**, etc., placing an **inoculum** into (or onto) medium to initiate a culture. 2. In immunology, to carry out **immunization**. 3. In plant pathology, application of **pathogen** spores etc. on to plants under conditions where infection should result in the absence of resistance.

inoculation cabinet Small room or cabinet for inoculation (of tissue or **microorganism** cultures) operations, often with a current of sterile air to carry contaminants away from the work area.

inoculum (pl.: inocula) 1. A small piece of tissue cut from **callus**, or an **explant** from a tissue or organ, or a small amount of cell material from a suspension culture, transferred into fresh medium for continued growth of the culture. *See*: **minimum inoculum size**. 2. Microbial spores or parts (such as mycelium). 3. **Vaccine**.

inorganic compound Historically, chemicals that could not be derived from living processes. In modern usage, chemicals that do not contain carbon, although carbonates and a few other simple carbon compounds are generally regarded as inorganic.

inositol A cyclic acid (hexahydroxycyclohexane) that is a constituent of certain cell phosphoglycerides. It is a nutrient frequently referred to as a "vitamin" in plant tissue culture. Also acts as a **growth factor** in some animals and microorganisms.

inositol lipid A membrane-anchored **phospholipid** that transduces hormonal signals by stimulating the release of any of several chemical messengers.

insecticide A substance that kills insects.

insert 1. To incorporate a **DNA** molecule into a **cloning vector**; also used as a noun to describe such a DNA molecule. 2. To introduce a gene or gene **construct** into a new genomic site or into a new genome.

insertion element Generic term for **DNA** sequences found in bacteria capable of genome insertion. Postulated to be responsible for **site-specific phage** and **plasmid** integration. *Synonym*: **insertion sequence**.

insertion mutation Changes in the base sequence of a **DNA** molecule resulting from the random integration of DNA from another source. *See*: **mutation**.

insertion sequence *See*: **insertion element**.

insertion site 1. A unique **restriction site** in a **vector DNA** molecule into which **foreign DNA** can readily be inserted. This is achieved by treating both the vector and the insert with the relevant **restriction.endonuclease** and then ligating the two different molecules, both having the same **sticky ends**. *Synonym*: **cloning site**. 2. The position of integration of a **transposon**.

***in-situ* conservation** The conservation of ecosystems and natural habitats and the maintenance and recovery of viable populations of species in their natural surroundings and, in the case of domesticated or cultivated species, in the surroundings where they have developed their distinctive properties.

instability A lack of consistent phenotype, usually as a result of uncontrolled genetic changes. These may be due to **transposon** activity, or in cell lines, to changes in **karyotype**.

insulin A peptide **hormone** secreted by the Langerhans islets of the pancreas, and that regulates the level of sugar in the blood.

integrating vector A vector that is designed to integrate cloned **DNA** into the host's chromosomal **DNA**.

integration The recombination process which inserts a small **DNA** molecule (usually by **homologous** recombination) into a larger one. If the molecules are circular, integration involves only a single **crossing-over**; if linear, then two crossings-over are required.

integration-excision region (Abbreviation: I/E). The portion of **bacteriophage** lambda (ë) **DNA** that enables ë-**DNA** to be inserted into a specific site in the ***E. coli*** bacteriophage lambda **chromosome** or excised from this site.

integument One of the layers that enclosed the ovule, and is the precursor of the **seed** coat.

intellectual property rights (Abbreviation: IPR). The legal framework, which includes patenting and **plant variety protection**, by which inventors control the commercial application of their work.

intensifying screen A plastic sheet impregnated with a rare-earth compound, such as calcium tungstate, which reacts to radiation by emitting light. When placed on one side of a piece of X-ray film with a radioactive sample on the other side, the intensifying screen will capture some of the radioactive energy which has passed through the film, exposing the X-ray film and so enhancing the sensitivity of the detection. Often used in **Southern** and **northern blotting** procedures.

intercalary 1. Meristematic tissue or growth not restricted to the **apex** of an organ, i.e. growth at nodes. 2. Referring to internal segments of a chromosomes (i.e. not at the ends).

intercalary growth A pattern of **stem** elongation typical of grasses. Elongation proceeds from the lower internodes to the upper internodes through the differentiation of meristematic **tissue** at the base of each internode.

intercalating agent A chemical capable of inserting between adjacent base pairs in a double-stranded nucleic acid. A prominent example is **ethidium bromide**.

intercellular space The pore space between cells, especially typical of leaf tissues.

interfascicular cambium Cambium that arises **between vascular bundles**.

interference The effect of one **crossing over** event in altering the probability of another **crossing over** event occurring at a nearby location. This **probability** can be either increased (positive interference) or decreased (negative interference), but the latter is the more usual.

interferon One of a group of small **proteins** synthesized by certain **T cells** of vertebrates, which inhibit **virus** replication. There are three types of interferon in humans. *See*: **cytokine**.

intergeneric cross A **hybrid** made between parents belonging to two different **genera**.

intergenic regions Non-coding **DNA** located between genes; this comprises a variable but considerable proportion of all eukaryotic genomic **DNA**, and its function is largely unknown.

intergenic spacer (Abbreviation: IGS). Non-coding **DNA** separating tandemly arranged copies of a repeated gene **sequence** (typically **ribosomal DNA**). Of particular interest because, unlike the **coding** sequence itself, the spacers show high levels of interspecific sequence polymorphism, and are thus useful as assays for species identification.

interleukin A group of **proteins** that transmit signals between immune cells and are necessary for mounting normal immune responses. *See*: **cytokine.**

internal guide sequence (Abbreviation: IGS). *See*: **guide sequence.**

internal transcribed spacer (Abbreviation: ITS). Non-coding regions separating the individual components of the **ribosomal DNA** units. These regions show much more sequence **polymorphism** than the genic regions themselves, and therefore, like the **intergenic spacers**, are useful a source of genetic markers for the ribosomal **DNA** locus.

International Treaty on Plant Genetic Resources for Food and Agriculture The international treaty resulting from the revision of the **International Undertaking on Plant Genetic Resources** was adopted by the 2001 FAO Conference as a binding international instrument to enter into force after ratification by 40 states. Its objectives are the conservation and sustainable use of plant genetic resources for food and agriculture and equitable sharing of the benefits of this use.

International Undertaking on Plant Genetic Resources The first comprehensive voluntary, international agreement (adopted in 1983) dealing with plant genetic resources for food and agriculture. Designed as an instrument to promote international harmony in matters regarding access to plant genetic resources for food and agriculture. Following extensive negotiations to revise the Undertaking in harmony with the Convention on Biological Diversity, the binding **International Treaty on**

Plant Genetic Resources for Food and Agriculture was adopted by the 2001 FAO Conference.

internode The region of a **stem** between two successive nodes.

interphase The stage in the **cell cycle** when the cell is not dividing and during part of which **DNA replication** occurs; it follows **telophase** of one mitotic division and extends to the beginning of **prophase** in the next division.

intersex Synonym of **hermaphrodite**.

inter-simple sequence repeat (Abbreviation: ISSR). A **PCR**-based molecular marker **assay** of genomic sequence lying between adjacent **microsatellites**. Primers carrying, at their 3'-end, sequence complementary to the repeat unit of the microsatellite will **amplify** this genomic **DNA**.

interspecific cross A **hybrid** made between parents belonging to two different species. *See*: **intrageneric cross, intraspecific cross**.

intervening sequence *See*: **intron**.

intracellular Occurring within a cell.

intracytoplasmic sperm injection (Abbreviation: ICSI). The **micro-injection** of a single sperm into the **cytoplasm** of a mature **oocyte**.

intrageneric cross A **hybrid** made between parents belonging to two species in the same **genus**. *See*: **intraspecific cross, interspecific cross.**

intrageneric Within a genus, such as an intrageneric **cross**, or intrageneric variation.

intragenic complementation Occurs when wild type phenotype is restored in an $\mathbf{F_1}$ individual made by crossing two independent mutants, carrying different **heteroalleles**.

intraspecific cross A **hybrid** made between parents belonging to the same **species**. *See*: **intrageneric cross, interspecific cross**.

intraspecific Within a species, such as an intraspecific **cross**, or intraspecific variation.

introgression The introduction of new alleles or gene(s) into a **population** from an exotic source, usually another species. This is achieved by repeated backcrossing of the initial **hybrid** in order to eliminate all genetic changes except for the desired new gene(s).

intron A segment of the primary **transcript** of a eukaryotic gene, removed (before the mature **mRNA** is translated) in a process known as intron **splicing**. Some eukaryotic genes contain a large number of introns, which make up the bulk of the **DNA** sequence of the gene. Introns are also found in genes whose **RNA** transcripts are not translated, namely eukaryotic **rRNA** and **tRNA** genes. In these cases the intron sequence does not appear in the functional **RNA** molecule. *Synonym*: **intervening sequence**.

invasiveness The ability of a plant, particularly a **weed**, to spread beyond its presently established site, and become established in new locations.

inversion A chromosome re-arrangement, which involves the re-orientation of a segment so that the order of a linear array of genes within it is reversed.

inverted repeat Two sequences of nucleotides occurring in one strand, where, relative to the first sequence, the second has **complementary** bases but in the inverted order. Under appropriate conditions this allows formation of a **hairpin loop** in the single strand. *See*: **palindrome**.

ion channel A **protein** integral to a **cell** membrane, through which selective ion transport occurs.

IPR Abbreviation for **intellectual property rights**.

IPTG Abbreviation for isopropyl-3-D-thiogalactopyranoside. A synthetic **inducer** of beta-**galactosidase** activity in many bacteria. Used in combination with the synthetic chromogenic **substrate** Xgal to differentiate recombinant from non-recombinant bacterial colonies in cloning strategies using **plasmid vectors** containing the *lacZ* gene: blue coloured colonies are produced when ?-galactosidase activity is not disrupted by an **insert**; but when it is disrupted, the colonies are white. Hence, white colonies are indicative of **recombinant** plasmids, and blue colonies of non-recombinant ones.

irradiation Illumination with electromagnetic radiation, typically of sufficiently high energy (low-wavelength UV or gamma, *etc.*) to disrupt biological macromolecules and hence induce **mutations**.

IS element Abbreviation for insertion sequence element. A short (800-1400 **nucleotide** pairs) **DNA sequence** found in bacteria that is capable of transposing to a new genomic location; **DNA** sequences contained within an IS element can be transposed along with the IS itself.

isoallele Multiple similar copies of a gene, usually located at independent positions in the genome, which **encode** similar gene products and produce the same, or a very similar phenotype. *See*: **allele**.

isochromosome A chromosome produced following an error in **meiosis**, in which the two arms are mirror images of each other. The presence of an isochromosome results in the **duplication** of all genes present on the originating chromosome arm.

isodiametric Commonly used to describe cells with equal diameters.

iso-electric focusing gel (Abbreviation: IEF gel). A variant of **gel electrophoresis**, in which **macromolecules** (usually **proteins**) are separated on the basis of differing iso-electric point, rather than on the basis of size.

isoenzyme *See*: **isozyme.**

isoform 1. A tissue-specific form of a **protein**. 2. Synonym of **isoenzyme.**

isogamy Fusion of gametes of similar size and structure.

isogenic A group of individuals that possesses the same genotype, irrespective of their being homozygous or heterozygous.

isogenic stock Strains of organisms that are genetically nearly identical, except with respect to identified genes. Generally produced by repeated backcrossing, or by **transformation.**

isolating mechanism The properties of an organism that prevent interbreeding (and therefore exchange of genetic material) between members of different species that inhabit the same geographical area.

isolation medium An optimum plant **tissue culture** medium suitable for **explant** survival, growth and development.

isomer 1. Structural isomers have the same chemical formula but different structures; e.g. leucine and isoleucine. 2. Stereoisomers are different topological forms of an otherwise single chemical structure, due to changes in bond configurations about some axis or plane of symmetry; eg, D- and L-glucose or cis- and trans-cinnamic acid.

isomerase Any of a class of enzymes that catalyse the re-arrangement of the atoms within a molecule, thereby converting one **isomer** into another.

iso-osmotic *See*: **isotonic**.

isotonic Solutions with the same osmotic potential, as a result of being of the same molar concentration. For **protoplasts** to avoid losing or gaining water, the medium they are suspended in must be isotonic with them. *See*: **hypertonic, hypotonic, osmosis**.

isotope One of two or more forms of an element that differ in the number of neutrons carried by the nucleus. Radioactive isotopes (radio-isotopes) are used as probes in many biochemical analyses.

isozyme A genetic **variant** of an **enzyme**. Isozymes for a given enzyme share the same function, but may differ in level of activity, as a result of minor differences in their **amino acid** sequence. Electrophoretic separation of isozymes has been used to distinguish between individuals and varieties.

ISSR Abbreviation for **inter-simple sequence repeat**.

ITS Abbreviation for **internal transcribed spacer**.

IVEP Abbreviation for ***in vitro* embryo production**.

IVF Abbreviation for ***in vitro* fertilization**.

IVM Abbreviation for ***in vitro* maturation**.

J

J *See*: **joining segment**.

Jiffy potä Pots made from wood pulp and peat, commonly used for transplanting **tissue culture**-derived plants into soil medium.

JIVET Abbreviation for **juvenile *in vitro* embryo technology**.

JIVT Abbreviation for **juvenile *in vitro* embryo technology**.

joining segment (Abbreviation: J). A small **DNA** segment that links genes in order to yield a functional **gene** encoding an **immunoglobulin**.

jumping gene *See*: **transposable element**.

jumping library *See*: **chromosome jumping**.

junk DNA *See*: **repetitive DNA**.

juvenile hormone A hormone secreted by insects from a pair of endocrine glands close to the brain. Its function is to inhibit metamorphosis so maintaining the larval features.

juvenile *in vitro* embryo technology (Abbreviation: JIVT or JIVET). A technology involving collection of immature eggs from young animals, their *in vitro* **maturation** and fertilization, and the transfer of the resultant embryos into recipient females. The method is designed to achieve rapid generation turnover.

juvenility Early phase of **development** in which an organism is incapable of sexual reproduction.

K

kanamycin An **antibiotic** of the aminoglycoside family that inhibits **translation** by binding to the **ribosomes**. Important as a substrate for selection of plant **transformants**.

kan^r Kanamycin-resistance gene. *See*: ***neo***^{r}, **selectable marker**.

kappa chain One of two classes of **antibody** light chains. The other is a lambda chain.

karyogamy The fusion of nuclei or nuclear material that occurs at fertilization during sexual reproduction.

karyogram A diagrammatic representation of the full **chromosome** set of a species, highlighting characteristic physical features of individual chromosomes.

karyokinesis The division of a cell nucleus. *See*: **meiosis**; **mitosis**.

karyotype The chromosome constitution of a cell, an individual, or of a related group of individuals, as defined both by the number and the morphology of the **chromosome**s, usually in mitotic metaphase; chromosomes arranged in order of length and according to position of **centromere**; also, the abbreviated formula for the chromosome constitution, such as 47, + 21 for human trisomy-21 (Down's syndrome).

kb Abbreviation for **kilobase** (of single-stranded nucleic acid).

kbp Abbreviation for **kilobase pairs** (of double-stranded **DNA**).

k_{cat}/K_m The catalytic efficiency of an enzyme-catalysed reaction. The greater the value of k_{cat}/K_m, the more rapidly and efficiently the **substrate** is converted into product.

k_{cat} The catalytic rate constant that characterizes an enzyme-catalysed reaction. The larger the k_{cat} value, the faster the conversion of **substrate** into product.

K_d Abbreviation for dissociation constant. Describes the strength of binding (or affinity) between molecules and their **ligands**. *See*: **avidity**.

kDa Abbreviation for kiloDalton. A unit of molecular mass equal to 1000 **Dalton**.

killer T cell T cells that kill cells displaying recognized **antigens**.

kilobase (Abbreviation: kb). A length of single-stranded **nucleic acid** composed of 1000 bases. One kilobase of **single-stranded DNA** has a mass of about 330 kiloDalton (exact mass depends on base composition).

kilobase pairs (Abbreviation: kbp). A length of double-stranded **DNA** composed of 1000 **base pairs**.

kinase An enzyme that catalyses the transfer of a phosphate group from a high energy state (as in ATP) to another molecule.

kinetics Dynamic processes involving motion. Often used as a suffix to indicate studies involving movement or rates of reactions. *See:* **pharmacokinetics, enzyme kinetics**.

kinetin A **cytokinin**.

kinetochore Structure at the **centromere** of eukaryotic chromosomes. The kinetochore consists of inner and outer electron dense plates and a central zone containing **repetitive DNA** elements. Kinetochores are involved in the control of chromosome movement in cell division.

kinetosome Granular cytoplasmic structure which forms the base of a **cilium** or flagellum. *Synonym*: basal body.

kinin A substance promoting cell division. In plant systems, the prefix cyto- has been added (**cytokinin**) to distinguish it from kinin in animal systems.

Klenow fragment A truncated form of **DNA polymerase** I from *E. coli*, used extensively for the production of synthetic **DNA** molecules as it retains **polymerase** and 3'-exonuclease activities, but not 5'-exonuclease activity.

K_m A dissociation constant that characterizes the binding of an enzyme to a substrate. The smaller the value of K_m, the tighter the binding of the **enzyme** to the **substrate**. Also called the Michaelis constant.

knockout A mutant individual, in which a single functional **gene** has been replaced by a non-functional form of the gene. Used to understand gene function via the comparison of the **phenotypes** of **wild type** and knockouts.

L

label A compound or atom that is attached to, or incorporated into, another molecule in order to allow detection of the latter's presence. Commonly, labels exploit radioactivity, fluorescence or antigenicity. *Synonym*: tag.

labelling The process of attaching or inserting a **label** into a molecule. Most often in the context of nucleic acids or **proteins**.

***lac* repressor-*lac* promoter system** *See*: **IPTG**.

lactose A disaccharide sugar produced in milk, composed of one unit each of glucose and galactose.

lag phase 1. The state of apparent inactivity preceding a response to a treatment; also called a latent phase. 2. The initial **growth phase**, during which **cell number** remains relatively constant, prior to the onset of rapid cell division.

lagging strand The strand of **DNA** that is synthesized discontinuously during **replication** (because **DNA** synthesis can proceed only in the 5'?3' direction). *See*: **Okazaki fragment**.

lambda chain One of two classes of **antibody**

light chains. The other is a kappa chain.

lambda phage A **bacteriophage** that infects *E. coli*, commonly used as a **cloning vector**. *See*: **integration-excision region**.

lamella A structure, plate or vesicle that is formed by two membranes lying parallel to each other.

lamina Blade or expanded part of a leaf.

laminar air-flow cabinet Cabinet designed for cell or **tissue culture** manipulations requiring a **sterile** environment. Achieved by a continuous, non-turbulent flow of filter-sterilized air over the working area. *Synonym*: **laminar air-flow hood**.

laminarin A storage **polysaccharide** of the brown algae.

lampbrush chromosome Large **diplotene** chromosomes present in **oocyte** nuclei, and particularly conspicuous in amphibians. These chromosomes have extended regions called loops, which are active sites of transcription. *See*: **diplonema**.

landrace In plant genetic resources, an early, cultivated form of a crop species, evolved from a wild population, and generally composed of a heterogeneous mixture of genotypes.

latent agent A **pathogen**, usually a virus, present in a host organism without producing any symptoms.

latent bud An inactive bud not held back by rest or dormant period, but which may start growth if stimulated.

latent phase *See*: **lag phase**.

lateral bud *See*: **axillary bud**.

lateral meristem A **meristem** giving rise to secondary plant tissues, such as the vascular and cork cambia.

lawn A uniform and uninterrupted layer of bacterial growth, typically on **agar** medium, in which individual colonies cannot be observed.

layering A technique for vegetative propagation, in which new plants produce **adventitious** roots before being severed from the parent plant.

LCR Abbreviation for **ligase chain reaction**.

LD_{50} Abbreviation for lethal dose$_{50\%}$. The amount of a substance required to kill 50% of the test population. The higher the LD_{50}, the lower the **toxicity** of the chemical in that specific test.

lead compound A chemical that has demonstrated promising biological activity in preliminary assays.

leader peptide *See*: **signal sequence.**

leader sequence A variable length sequence of **nucleotides** at the 5' end of an **mRNA** molecule that precedes the AUG **initiation codon** where **translation** begins and is not itself translated into **protein.**

leading strand The strand of **DNA** that is synthesized continuously during replication.

leaf blade The usually flattened portion of the leaf.

leaf bud cutting A **cutting** that includes a short section of **stem** with attached leaf.

leaf margin The edge of a leaf.

leaf primordium A lateral outgrowth from the **apical meristem**, which will become a leaf when fully developed and expanded.

leaf roll A symptom of some **virus** diseases, characterized by curling of the leaves. Can also occur as a response to water stress.

leaf scar Mark left on a **stem** after leaf abscission.

leaflet Expanded leaf-like part of a compound leaf.

leaky mutant A mutant in which the gene product still retains some biological activity.

lectin A group of plant **proteins** that can bind to specific **oligosaccharides** on the surface of cells, causing the cells to clump together.

leptonema Stage in **meiosis** immediately preceding **synapsis** and post **DNA** replication, in which the chromosomes appear as single, fine, threadlike structures.

leptotene (adj.) *See*: **leptonema.**

lethal allele A mutant form of a gene that, in the **homozygous** state, is fatal.

lethal gene *See*: **lethal allele.**

lethal mutation *See*: **lethal allele.**

leukocyte White blood cell, up to 0.02 mm in

diameter, of which there are normally 4-11 million per millilitre of human blood. There are several kinds, all involved in the body's defence mechanisms. Granulocytes have granules in their cytoplasm; monocytes ingest and feed on bacteria and other **microorganisms** that cause infection; **lymphocytes** include the **B cells** that are involved with the production of **antibodies**.

library A collection of cells, usually bacteria or yeast, that have been transformed with **recombinant** vectors carrying **DNA** derived from an unrelated **organism**. *See*: **cDNA library, expression library, genomic library**.

life cycle The sequence of events from a given developmental stage in one generation to the same stage in the following generation. In sexually reproduced organisms, the starting point is the fusion of gametes to form the zygote.

ligand A small molecule (e.g. activators, substrates and inhibitors of **enzyme** activity) bound to a **protein** by non-covalent forces; an ion or a molecule that binds to another chemical entity to form a larger complex.

ligase chain reaction (Abbreviation: LCR). A technique for the detection and **amplification** of target **DNA** sequences. Two **oligonucleotides** are synthesized which between them are **complementary** to the entire **target** sequence, one to the 5'-side and one to the 3'-side. If the target sequence is present in the **DNA** sample under examination, the oligonucleotides will bind to it with their ends abutting in the centre, and a heat-stable ligase will join them into a complete polynucleotide. No ligation occurs if the target sequence is absent or if the match between synthetic oligonucleotides and target sequence is imperfect in the region where they abut. At a high temperature, the new **polynucleotide** dissociates from the original **DNA** template, and upon cooling it and the original **DNA** serve as templates for a second cycle of hybridization, ligation and thermal dissociation. At each cycle, there is a doubling of the number of new complete polynucleotides.

ligase *See*: **DNA ligase**.

ligate, ligation The joining of two linear **double-stranded DNA** fragments by the formation of phosphodiester bonds.

lignification The thickening and strengthening of a plant **cell wall** with **lignin**.

lignin A group of high-molecular-weight amorphous polymers of phenylpropanoid compounds, giving strength to certain tissues. A major component of wood.

lignocellulose The combination of **lignin**, **hemicellulose** and **cellulose** that forms the structural framework of plant cell walls.

LINE Abbreviation for **long interspersed nuclear element**.

lineage A group of individuals, related by common descent, e.g. an *in vitro* **cell line** derived from a single cell.

linear phase The **growth phase** during cell culture when cell number increases arithmetically. The linear phase follows a period of exponential growth.

linearized vector A covalently closed circular **DNA vector** (typically a **plasmid**) which has been opened by restriction digestion to convert it to a linear molecule. In molecular cloning, **DNA** to be cloned is mixed with the linearized vector, and treated with **ligase** to join and recircularize the resulting hybrid molecule.

linkage disequilibrium *See*: **gametic phase disequilibrium**.

linkage equilibrium *See*: **gametic phase equilibrium**.

linkage map A linear or circular diagram that shows the relative positions of genes on a **chromosome** as determined by recombination fraction. *See*: **genetic map**.

linkage The tendency of a set of genes to be inherited together more often than would be expected if they were assorting independently. exists between two genes when they are located sufficiently close to one another on the same **chromosome** that a proportion of gametes is produced without crossing-over occurring between them.

linked gene, linked marker A gene or marker that is linked to another gene or marker.

linker A synthetic double-stranded **oligonucleotide** that carries the recognition sequence for one or more **restriction endonucleases**. The **ligation** of a linker to each end of a **DNA** fragment facilitates the preparation of the fragment for cloning into a **vector**. *See*: **polylinker**.

lipase A class of enzymes which break down lipids into their component fatty acids and glycerol. Lipases used in **biotechnology** are generally digestive, with a role in the breakdown of fats in food into their components, so that these can be used to make other materials.

lipid Any of a group of fats or fat-like compounds insoluble in water and soluble in fat solvents.

lipofection Delivery into eukaryotic cells of **DNA, RNA** or other compounds that have been encapsulated in **liposomes**.

lipopolysaccharide (Abbreviation: LPS). A compound containing **lipid** bound to a **polysaccharide**; often a component of microbial cells walls.

liposome A synthetic microscopic spherical structure consisting of a **phospholipid** bilayer membrane containing a user-defined aqueous solution. Liposomes can be used to transport relatively toxic drugs into diseased cells, where they can exert their maximum effect. **DNA** molecules may be entrapped in, or bound to the surface of, the vesicles, and subsequent fusion of the liposome with the **cell membrane** will deliver the **DNA** into the cell. Liposomes have been used to develop an efficient **transfection** procedure for *Streptomyces* bacteria.

liquefaction Enzymatic digestion (often by alpha-amylase) of gelatinized **starch** to form lower molecular weight **polysaccharides**.

liquid medium Culture solution, without a solidifying agent, for *in vitro* cell growth.

liquid membrane Thin films made up of liquids

(as opposed to solids) which are stable in another liquid (usually water). Thus, the liquid must not dissolve in the water, but nevertheless must be prevented from collapsing into small droplets.

liquid nitrogen Nitrogen gas condensed to a liquid with a boiling point of about -196 °C. Commonly used as a medium for long-term storage of biological materials. *See*: **cryopreservation**.

litmus paper A **pH** indicator paper. It turns red in acidic and blue in alkaline solutions.

live recombinant vaccine A **vaccine** created by the expression of a **pathogen antigen** in a non-pathogenic organism.

live vaccine A living, non-virulent form of a pathogenic **microorganism** or **virus** used to elicit an **antibody** response for the protection against **infection** by a virulent form of the same **pathogen**.

living modified organism (Abbreviation: LMO). "Living organism that possess a novel combination of genetic material obtained through the use of **modern biotechnology**" (Convention on Biological Diversity). Synonym of **GMO**, but restricted to organisms that can endanger biological diversity.

LMO Abbreviation for **living modified organism**.

locus (pl.: loci) A site on a chromosome.

lod score The logarithm of the odds of **linkage** between two loci. Used to measure the statistical support for linkage.

log phase Abbreviation for logarithmic phase.

logarithmic phase The **growth phase** in cell **culture**, during which **cell number** doubles every 20-30 minutes. *Synonym*: exponential phase.

long interspersed nuclear element (Abbreviation: LINE). Families of common **DNA** elements, of average length 6.5 kb, which are dispersed at numerous locations within the genome. The human **genome** contains over 500,000 LINEs (representing ca. 16% of the genome). They appear to represent degenerate copies of transposable elements. *See*: **SINE**.

long template A **DNA** strand, synthesized during **PCR**, which has a **primer** sequence at one end but is extended beyond the site that is **complementary** to the second primer at the other end.

long terminal repeat (Abbreviation: LTR). A characteristic sequence of **nucleotides** that occurs at each end of a **retrovirus** element that has become integrated into the host **genome**. Involved in the integration process.

long-day plant Plants requiring a period of short nights before the switch from vegetative to reproductive growth can be initiated. *See*: **Short-day plant.**

loop bioreactor Fermenters in which material is cycled between a bulk tank and a smaller tank or loop of pipes. The circulation helps to mix the materials and to ensure that gas injected into the **fermenter** is well distributed in the liquid. Particularly useful for **photosynthetic** fermentations, where the photosynthesizing organisms are passed through a system of many small transparent pipes, which allow the access of light.

LPS Abbreviation for **lipopolysaccharide**.

LTR Abbreviation for **long terminal repeat.**

luteinizing hormone A pituitary hormone which causes growth of the yellow body of the **ovary** and also stimulates activity of the interstitial cells of the testis.

luxury consumption Nutrient absorption by an organism in excess of that required for optimum growth and productivity.

lyase Any of a class of enzymes that catalyse either the cleavage of a double bond and the addition of new groups to a substrate, or the formation of a double bond.

lymphocyte White blood cells that are important components of the immune system of vertebrates. *See*: **B cell, T cell.**

lymphokine Generic name for **proteins** that are released by **lymphocytes** to act on other cells involved in the immune response. The term includes **interleukins** and **interferons**. A subclass of **cytokines**. *See*: **monokine.**

lymphoma Cancer originating in the lymph

nodes, spleen and other lympho-reticular sites.

lyophilize *See*: **freeze-dry**ing.

lysis The destruction or breakage of cells either by viruses or by chemical or physical treatment.

lysogen A bacterial cell whose **chromosome** contains integrated **bacteriophage DNA**.

lysogenic Bacteria or bacteriophages undergoing **lysogeny**.

lysogenic bacterium Bacterium harbouring temperate (non-virulent, **lysogenic**) bacteriophages.

lysogeny A condition in which a **bacteriophage** genome (**pro-phage**) survives within a host bacterium, either as part of the host chromosome or as part of an **extrachromosomal** element, and does not initiate **lysis**.

lysosome A membrane-bound sac within the **cytoplasm** of animal cells that contains enzymes responsible for the digestion of material in food vacuoles, the dissolution of foreign particles entering the **cell** and, on the death of the cell, the breaking down of all cell structures. The digestive system of the cell.

lysozyme A naturally occurring **enzyme** extracted from egg white **protein** and other animal and plant sources, which attacks the cell wall of gram-positive bacteria leading to cell **lysis** and death.

lytic A phase of the virus life cycle during which the **virus** replicates within the host cell, releasing a new generation of viruses when the infected cell undergoes **lysis**.

lytic cycle The steps in viral production that lead to cell **lysis**.

M13 A single-stranded **DNA bacteriophage** used as a **vector** for **DNA** sequencing.

M13 strand The single-stranded **DNA** molecule that is present in the infective form of **bacteriophage M13**.

MAAP Abbreviation for **multiple arbitrary amplicon profiling**.

mAb Abbreviation for **monoclonal antibody**.

macerate To disintegrate tissue to disrupt cells. Commonly achieved via mechanical shearing, **plasmolysis** or enzymatic **cell wall** degradation.

macromolecule Any high molecular weight molecule. Often used as a synonym for **polymers**.

macronutrient A major chemical element essential for normal growth and development. In **tissue culture** media, macronutrients are those required in concentrations above 0.5 millimole/litre.

macrophage Large white blood cells that ingest foreign substances and display on their surfaces **antigens** which are recognized by other cells of the immune system.

macropropagation Production of plant clones from growing parts.

macrospore *See*: **megaspore**.

mad cow disease Colloquial term for **bovine spongiform encephalopathy**. *See* **proteinaceous infectious particle**

MADS-box A highly conserved **DNA** sequence **motif** found in a large family of **transcription factors**, most of which play important roles in developmental processes. Most prominently, the MADS box genes known in flowering plants are intimately involved in the control of flower morphogenesis.

magenta A type of plastic container frequently used for plant **micropropagation** and **tissue culture**.

major histocompatibility antigen A cell-surface **protein** or **glycoprotein** that allows the immune system to distinguish foreign or "non-self" from "self". A better term is **histoglobulin**. These are the **antigens** that must be matched between donors and recipients during **organ** and **tissue** transplants to prevent rejection.

major histocompatibility complex (Abbreviation: MHC). The large cluster of genes that encode the **major histocompatibility antigens** in mammals.

malt extract A mixture of organic compounds prepared from malt, used as a **culture medium** additive.

malting Enzymatic reduction of **starch** to sugars in germinating grain, used in brewing.

mammary gland The milk-producing **organ** of female mammals.

management of farm animal genetic resources The sum total of technical, policy and logistical operations involved in understanding (characterization), using and developing (utilization), maintaining (conservation), accessing, and sharing the benefits of animal genetic resources.

mannitol A sugar alcohol widely distributed in plants. Commonly used as a nutrient and **osmoticum** in suspension media for plant **pro-**

toplasts.

mannose A hexose component of many polysaccharides, occasionally used as a **carbohydrate** source in plant **tissue culture** media.

map 1. verb: to determine the relative positions of loci (genes or **DNA** sequences) on a **chromosome. Linkage** maps are obtained from the frequency of **recombination** between loci. Physical maps are obtained commonly by the use of *in situ* hybridization of cloned **DNA** fragments to **metaphase** chromosomes, or by **somatic-cell hybrids** or **radiation hybrids**. 2. noun: a diagram showing the relative position of, and distances between, loci on a chromosome.

map distance The standard measure of **genetic distance** between loci, expressed in **centiMorgans** (cM) or **map units**. Estimated from **recombination** fraction via a **mapping function**. For small recombination fractions, map distance in cM equals the **recombination fraction** in %.

map unit One **centiMorgan** (1 cM) *See*: **map distance; crossing-over unit.**

mapping function A mathematical expression relating observed **recombination fraction** to **map distance**.

mapping The construction of a localized (around a gene), or broad-based (whole **genome**) genetic map. More generally, determining the location of a locus (gene or genetic marker) on a chromosome.

mariculture *See*: **aquaculture.**

marker An identifiable **DNA** sequence that is inherited in Mendelian fashion, and which facilitates the study of **inheritance** of a **trait** or a linked gene.

marker gene A gene of known function or known location, used for **marker-assisted selection** or genetic studies.

marker peptide A portion of **fusion protein** that facilitates its identification or purification.

marker-assisted introgression The use of **DNA** markers to increase the speed and efficiency of **introgression** of a new allele(s) or gene(s) into a breeding population. The markers will be closely linked to the **gene**(s) in question.

marker-assisted selection (Abbreviation: MAS). The use of **DNA** markers to improve response to selection in a population. The markers will be closely linked to one or more target loci, which may often be **quantitative trait loci**.

MAS Abbreviation for **marker-assisted selection**.

mass selection As practised in plant and animal breeding, the selection of a number of individuals, on the basis of their individual **phenotypes**, to interbreed to form the next generation.

maternal effect An effect attributable to a genetic contribution of the female parent of the individual being evaluated.

maternal inheritance Inheritance controlled by non-nuclear genes (e.g. **mitochrondria, chloroplast**) that are transmitted only through the female line.

matric potential A **water potential** component, always of negative value, resulting from the presence of solid (often finely divided) surfaces; primarily responsible for water uptake by a dry **seed** prior to germination.

maturation The formation of gametes or spores.

MCS Abbreviation for **multiple cloning site**. *See*: **polylinker**.

MDA Abbreviation for multiple drop array. *See*: **microdroplet array**.

mean In statistics, the arithmetic average; the sum of all measurements or values in a sample divided by the sample size.

media *See*: **culture medium; medium**.

median In a set of measurements, the central value above and below which there are an equal number of measurements.

medium (pl.: media) 1. In plant tissue culture, a term for the liquid or solid formulation upon which plant cells, tissues or organs develop. *See*: **culture medium**. 2. In general terms, a **substrate** for plant growth, such as nutrient solution, soil, sand, etc., e.g. potting medium.

medium formulation In tissue culture, the particular constituents for the culture medium, commonly comprising macro- and micro-elements, **vitamins**, plant **hormones**, and a **carbohydrate** source. Some formulations are very specific to the kind of **explant** or plant species that can be maintained; some are very general.

mega yeast artificial chromosome A **yeast artificial chromosome** (YAC) which can carry particularly large inserts (up to 1Mbp) - standard YACs typically carry inserts of up to 500kbp.

megabase (Abbreviation: Mb). A length of **DNA** consisting of 10^6 **bases**.

megabase cloning The **cloning** of large **DNA** fragments of the order of 1Mb.

megaDalton (Abbreviation: MDa). One megaDalton is equal to 10^6 **Dalton**.

megagametophyte The female gametophyte; the plant that develops from a **megaspore**.

megaspore The female **gametophyte** in heterosporous plants. *Synonym*: **macrospore**.

meiosis The two-stage process in sexual reproduction by which the **chromosome** number is reduced from the **somatic** to the **haploid** number. The first division, in which **homologous** chromosomes pair and exchange genetic material, is followed by amitotic division. The nucleus divides twice, but the chromosomes only once, generating haploid nuclei, which develop into the **gametes** (**egg** and **sperm** in animals; **egg** and **s** in plants).

meiotic analysis The use of patterns of **chromosome pairing** at meiotic **prophase** and **metaphase** to detect relationships between chromosomes, from which can be deduced the relationship between the parents of the organism studied.

meiotic drive Any mechanism that causes a particular **allele** or **chromosome** to be over-represented in a **population** of gametes.

meiotic product *See*: **gamete**.

melanin Dark pigment, produced by specialized epidermal cells called melanocytes.

melting temperature (Abbreviation: T_m). The temperature at which a **double-stranded DNA** molecule denatures into separate single strands. T_m is determined by the length of the molecule and its base composition. **DNAs** rich in G:C base pairs have higher T_m than A:T rich **DNA**, because since three hydrogen bonds are formed between G and C, but only two between A and T.

membrane bioreactor A vessel in which cells are cultured on or behind a permeable membrane, which allows the diffusion of nutrients to the cells, but retains the cells themselves. A variation is the hollow-fibre reactor.

memory cell Long-lived **B cells** and **T cells** that mediate rapid secondary immune responses to a previously encountered **antigen**.

Mendelian population A natural, interbreeding unit of sexually reproducing plants or animals sharing a common **gene** pool.

Mendelian segregation Occurs when **alleles** are inherited according to **Mendel's Laws**.

Mendel's Laws Two laws summarizing Gregor Mendel's theory of inheritance. The Law of **Segregation** states that each hereditary characteristic is controlled by two 'factors' (now called **alleles**), which segregate and pass into separate germ cells. The Law of Independent **Assortment** states that pairs of 'factors' segregate independently of each other when germ cells are formed. *See*: **independent assortment**; **linkage**.

mericlinal Refers to a **chimera** with tissue of one **genotype** partly surrounded by that of another genotype.

mericloning A propagation method using shoot tips in culture to proliferate multiple buds, which can then be separated, rooted and planted out.

meristele The branch of a **stele** supplying the leaf.

meristem culture A tissue culture containing meristematic dome tissue without adjacent leaf **primordia** or stem tissue. The term may also imply the culture of meristemoidal regions of plants, or meristematic growth in culture.

meristem tip An **explant** comprising the **meristem** (meristematic dome) and usually one pair of leaf **primordia**. Also refers to explants originating from **apical meristem** tip or lateral or axillary meristem tip.

meristem tip culture Cultures derived from **meristem** tip **explants**. Used widely to achieve **virus** elimination and axillary shoot proliferation, less commonly for callus production.

meristem Undifferentiated but determined plant tissue, in which the cells are capable of active division and **differentiation** into specialized tissues such as shoots and roots.

meristemoid A localized group of **callus** cells, characterized by their accumulation of starch, **RNA** and **protein**, and giving rise to **adventitious** shoots or roots.

merozygote Partial **zygote** produced by a process of partial genetic exchange, such as **transformation** in bacteria.

mesh bioreactor *See*: **filter bioreactor**.

mesoderm The middle germ layer that forms in the early animal **embryo** and gives rise to parts such as bone and connective tissue.

mesophile A microorganism able to grow in the temperature range 20-50 °C; optimal growth often occurs at about 37 °C. *See*: **psychrophile**, **thermophile**.

mesophyll Leaf parenchyma **tissue** found between epidermal layers.

messenger RNA Abbreviation: **mRNA**.

metabolic cell A cell that is not dividing. *See:* **metabolism**.

metabolism The biochemical processes whereby nutritive material is converted to living matter, or aids in building living matter, or by which complex substances and food are broken down into simple substances.

metabolite A low-molecular-weight biological compound that is usually synthesized enzymically.

metabolomics The large-scale study of the full complement of **secondary metabolites** produced by a given species in all its tissues and growth stages.

metacentric chromosome A chromosome in which the **centromere** is located in the middle and, consequently, the chromosome arms are of about equal length.

metal affinity chromatography A chromatographic technique, in which a compound interacting with a specific metal ion can be captured by immobilizing the relevant ion on the column's solid matrix.

metalloenzyme An enzyme which requires the presence of a metal in order to be catalytically active.

metallothionein A protective **protein** that binds heavy metals such as cadmium and lead.

metaphase Stage of **mitosis** or **meiosis** (following **prophase** and preceding **anaphase**) during which the **chromosomes**, or at least the **kinetochores**, lie in the central plane of the spindle. The stage of maximum chromosome condensation, at which **karyotypes** are generally described. In the first division of meiosis, metaphase represents the stage at which **meiotic analysis** is generally performed.

metastasis The spread of cancer cells to previously unaffected organs.

methylation The addition of a methyl group (-CH_3) to a molecule, most commonly in the context of **DNA** where **cytosine** and, less often, **adenine** residues can be modified in this way, sometimes resulting in a change in **transcription**. *See*: **epigenetic variation**.

MHC Abbreviation for **major histocompatibility complex**.

Michaelis constant *See*: K_m.

microalgal culture Culture in **bioreactors** of microalgae (including seaweeds).

micro-array A large set of cloned **DNA** molecules immobilized as a compact and orderly pattern of sub-microlitre spots onto a solid matrix (typically a glass slide). Used to analyse patterns of gene expression, presence of markers, or **nucleotide** sequence. The major advantage of micro-arrays is the extent to which the process of genotyping can be automated, thereby enabling large numbers of individuals to be simultaneously genotyped at many loci. A similar approach may be used with other immobilized components for other purposes. *Synonym*: **chip** or **DNA chip**. *See:* **somatic cell hybrid panel, radiation hybrid cell**

panel

microbe *See*: **microorganism**.

microbial mat Layered microbial populations, usually growing on the surface of a **solid medium** or on a membrane.

microbody A frequently spherical cellular organelle, bound by a single membrane, 20-60 nm in diameter, and containing a variety of enzymes.

micro-carrier Small particles used as a support material for (particularly mammalian) cells, which are too fragile to be pumped and stirred as bacterial cells are in a large-scale culture.

microdroplet array (Abbreviation: MDA). A technique used to simultaneously evaluate large numbers of media modifications, employing small quantities of medium into which are placed small numbers of cells or **protoplasts**. These form a **monolayer** at the droplet meniscus and can easily be examined. *Synonyms*: multiple drop array, hanging droplet technique

micro-element A nutritional element required in very small quantities.

micro-encapsulation A process of enclosing a substance in very small sealed capsules from which material is released by heat, solution or other means.

micro-environment A small-scale environment in which the conditions (temperature, humidity, **pH** etc.) are distinct; typically used in connection with the surroundings of a living object.

microfibril Microscopic fibres visible only at the high magnification of the electron microscope.

microgametophyte *See*: **pollen**.

micrograft *See*: **shoot-tip graft**.

micro-injection The introduction of small amounts of (usually) liquid material (**DNA**, **RNA**, **enzymes**, cytotoxic agents) into a defined **tissue** or single cell with a fine, microscopic needle.

micro-isolating system Mechanical separation of single cells or protoplasts thus allowing them to proliferate individually.

micronucleus A **nucleus**, distinct from and

smaller than the main **nucleus**, but lying within the same cell. They usually arise following abnormal meiotic or mitotic **telophases**, where individual **chromosomes** or chromosome fragments do not reach the pole.

micronutrient An essential element normally required in **cell culture** at concentrations under 0.5 millimole/litre.

microorganism Organism visible only under magnification.

microplast Vesicle produced by subdivision and fragmentation of protoplasts or thin-walled cells.

microprojectile bombardment *See*: **biolistics**.

micropropagation Miniaturized ***in vitro*** multiplication and/or **regeneration** of plant material under **aseptic** and controlled environmental conditions.

micropyle 1. A small opening in the surface of a plant **ovule** through which the **pollen** tube passes prior to fertilization. 2. A small pore in some animal cells or tissues.

microsatellite A segment of **DNA** characterized by a variable number of copies (typically 5-50) of a sequence of around 5 or fewer bases (called a **repeat unit**). At any one locus (genomic site), there are usually several different "**alleles**" in a population, each allele identifiable according to the number of repeat units. This existence of **multiple alleles** (high level of **polymorphism**) has enabled microsatellites to be developed as powerful markers in many different species. They are detected by the **polymerase chain reaction**.

microspore The immature male **gametophyte** in **seed** plants; the **haploid** male cell that ripens into a **pollen** grain.

microtuber Miniature tuber, produced in **tissue culture**, which is readily regenerable into a normal tuberous plant.

microtubule Self-assembling components of the cytoskeleton. Microtubules are cylindrical **protein** polymers, interconnected by cross-bridging proteins, which structurally and dynamically organize functional activities in living cells. They form the **spindle** during mitosis.

middle lamella A thin exclusively plant membrane separating two adjacent **protoplasts** and remaining as a distinct cementing layer between adjacent **cell walls**.

mid-parent value The average of the phenotypic measure, with respect to a given trait, of the two parents used to generate the **population** being analysed.

mineralization The conversion of organic compounds into inorganic (mineral) ones. For example, the conversion of **ethanol** into carbon dioxide and water.

minimum effective cell density The cell density below which reproducible cell growth fails. The minimum density is determined by both the source function of the **tissue** (**species**, **explant**, **cell line**) and by the culture phase of the initial **inoculum**.

minimum inoculum size The critical volume of **inoculum** required to initiate culture growth, due to the diffusive loss of cell materials into the medium. The subsequent culture growth cycle is dependent on the inoculum size, which is determined by the volume of **medium** and size of the culture vessel.

mini-prep A small-scale preparation of **plasmid** or **phage DNA**. Used to purify cloned **DNA** from the **DNA** of the vector.

minisatellite A form of **variable number tandem repeats** in which the **repeat unit** size ranges from 10-100 bp. Used for **DNA fingerprinting** following **Southern hybridization**. Generally concentrated at the ends of chromosomes and in regions with a high frequency of **recombination**.

minituber Small tubers (5-15 mm in diameter) formed on shoot cultures or cuttings of tuber-forming crops, such as potato.

mismatch repair A **DNA repair** process that corrects mismatched base pairs.

mismatch The occurrence of a non-complementary pairs of bases in a **double helix** of **DNA**, e.g. A:C, G:T.

missense mutation A mutation that changes a **codon** for one **amino acid** into a codon specifying another amino acid.

mist propagation Application of fine droplets of water to maintain humidity around **plantlets** or cuttings, which have not yet developed effective roots.

mite Free-living and parasitic arachnids. Infestation of plant crops reduces yield through the destruction of leaf tissue. Also can infest plant **tissue culture** work areas, contaminating culture vessels, thereby spreading bacteria and fungi.

mitochondrial DNA (Abbreviation: mtDNA). A circular **DNA** found in mitochondria. In mammals, mtDNA makes up less than 1% of the total DNA, but in plants the amount is variable. It encodes **rRNA** and **tRNA** and some mitochondrial **proteins** (up to 30 in animals).

mitochondrion (pl.: mitochondria) Organelle possessing its own **DNA** which appear in all eukaryotic cells (and never in prokaryotic cells) and produce **adenosine triphosphate** as an energy source for the cell via oxidative phosphorylation. Mitochondria contain many enzymes of the respiratory cycle, although most of these **proteins** are nuclear encoded.

mitogen A substance that can cause cells to initiate mitosis.

mitosis Splitting of replicated chromosomes, and the division of the **cytoplasm** to produce two genetically identical daughter cells. On the basis of the appearance of the **chromosomes**, it is separated into five stages: **interphase, prophase, metaphase, anaphase** and **telophase**.

mixed bud A bud containing both rudimentary leaves and flowers.

mixoploid Groups of cells with variable (a mix of **euploid, aneuploid, polyploid**) **chromosome** numbers.

mobilization 1. The transfer between bacteria of a non-conjugative **plasmid** by a conjugative plasmid. 2. The transfer between bacteria of chromosomal genes by a conjugative plasmid.

mobilizing function The genes on a **plasmid**

that give it the ability to facilitate the transfer of either a non-conjugative or a conjugative plasmid from one bacterium to another.

mode In a frequency distribution, the class having the greatest frequency.

model 1. A mathematical description of a biological phenomenon. 2. A simplified biological system used to test hypotheses (e.g. ***Arabidopsis*** *thaliana* as a model plant).

modern biotechnology The application of: a. In vitro nucleic acid techniques, including recombinant deoxyribonucleic acid (**DNA**) and direct injection of nucleic acid into cells or organelles, or b. Fusion of cells beyond the taxonomic family, that overcome natural physiological reproductive or recombination barriers and that are not techniques used in traditional breeding and selection (Convention on Biological Diversity).

modification Enzymatic attachment of one or more chemical groups to a **macromolecule**, affecting its biological activity or properties. *See*: **methylation**, **glycosylation**, **phosphorylation**.

modifying gene A gene that affects the expression of some other gene.

MOET Abbreviation for **multiple ovulation and embryo transfer**.

molecular biology The study of living processes at the molecular level.

molecular chaperone *See*: **chaperone**.

molecular cloning The biological **amplification** of a **DNA** sequence via the mitotic division of a host cell into which it has been transformed or transfected. *See*: **cloning**.

molecular genetics The study of the expression, regulation and inheritance of genes at the level of **DNA** and its **transcription** products.

molecular marker A genetic marker which is assayed at the **DNA** level.

molecular pharming *See*: **biopharming**.

molecule The stable union of two or more atoms; some organic molecules contain very large numbers of atoms.

monoclonal antibody (Abbreviation: mAb). An **antibody**, produced by a **hybridoma**, directed against a single **antigenic determinant** of an antigen.

monocot Abbreviation for **monocotyledon**.

monocotyledon (Abbreviation: monocot). A flowering plant whose **embryo** has one **cotyledon**. Examples are cereals (corn, wheat, rice etc.), banana, and lily.

monoculture The agricultural practice of cultivating a single crop over a whole farm or area.

monoecious A plant species that has separate male and female flowers on the same plant (e.g. maize).

monogastric animal A non-ruminant animal with a simple stomach.

monogenic Trait controlled by a single gene. *Opposite*: multigenic, **polygenic**.

monohybrid cross A cross between parents differing in only one trait or in which only one **trait** is being considered.

monohybrid Heterozygous with respect to one gene.

monokine Generic name for **proteins** that are released by monocytes to act on other cells involved in the immune response. A sub-class of **cytokines**.

monolayer A single layer of cells growing on a surface.

monolignols The building blocks of **lignin** that undergo polymerization.

monomer A small molecule (in the biological sciences typically individual amino acids, nucleotides or monosaccharides) that can combine with identical or similar others to form a larger, more complex molecule called a **polymer**.

monomorphic Absence of variation for a **marker**, **gene**, **chromosome**, or genetically determined **trait** in a **population**.

monophyletic A group of organisms that are assumed to have originated from the same ancestor.

monoploid *See*: **haploid**.

monosaccharide A simple sugar (e.g. glucose, fructose). *See*: **disaccharide, polysaccharide**.

monosomic A form of aneuploidy in which a **diploid** organism lacks one member of a **ho-**

mologous chromosome pair.

mono-unsaturates Oils containing mono-unsaturated fatty acids (i.e. where one $-CH_2-CH_2-$ group in the hydrocarbon chain is replaced by -CH=CH-).

monozygotic twin One of a pair of twins derived from a single fertilized egg. *Synonym*: **identical twin**.

morphogen A substance that stimulates the **development** of form or structure in an organism.

morphogenesis The development, through growth and differentiation, of form and structure in an organism.

morphogenic response The effect on the developmental history of a plant or its parts exposed to a given set of growth conditions or to a change in the environment.

morphology Shape, form, external structure or arrangement.

mosaic An organism or part of an organism that is composed of cells with different origin.

mother plant *See*: **donor plant**.

motif A **conserved sequence** of nucleotides or amino acids that can be associated with some function of, respectively, a length of **DNA** or a **protein**.

movable genetic element *See*: **transposon**.

mRNA Abbreviation for **messenger RNA**. The **RNA** molecule resulting from **transcription** of a **protein**-encoding gene, following any **splicing** (1). The information encoded in the mRNA molecule is translated into a gene product by the **ribosomes**.

MRU Abbreviation for minimum recognition units. *See*: **dAb**.

mtDNA Abbreviation for **mitochondrial DNA**.

multi-copy Describing **plasmids** which replicate to produce many copies per host bacterial cell.

multigene family A set of genes (not necessarily **mapping** to the same genomic location) that are related in **nucleotide** sequence and/or that produce **polypeptides** with similar **amino acid** sequences. Sequence similarity does not always result in functional similarity.

multigenic Trait controlled by several genes, as opposed to **monogenic**. *Synonym*: polygenic.

multi-locus probe A **DNA** sequence that hybridizes to a number of different genomic sites.

multimer A **protein** made up of more than one **polypeptide** chain.

multiple alleles The existence of more than two **alleles** at a locus in a population.

multiple arbitrary amplicon profiling A collective term for a number of related **polymerase chain reaction** techniques, all of which use arbitrary primers, and which generate a number of distinct **amplification** products. *See*: **random amplified polymorphic DNA**.

multiple cloning site (Abbreviation: MCS). *See*: **polylinker**.

multiple drop array (Abbreviation: MDA). *See*: **microdroplet array**.

multiple ovulation and embryo transfer (Abbreviation: MOET). A technology by which a single female that usually produces only one or two **offspring** can produce a litter of offspring. Involves stimulation of a female to shed large numbers of ova; natural mating or artificial insemination; collection of fertilized ova (either surgically, or non-surgically through the cervix); and transfer (usually non-surgical, through the cervix) of these fertilized ova to recipient females.

multiplex 1. The simultaneous **amplification** of a number of **amplicons** in a single **polymerase chain reaction**, achieved by including more than one set of **primers** in the reaction mix. 2. The inheritance pattern of alleles in **autopolyploids**. *See*: **quadruplex**.

multivalent vaccine A **vaccine** designed to elicit an **immune response** either to more than one **infectious agent** or to several different **antigenic determinants** of a single agent.

mutable gene A gene which has an unusually high rate of mutation.

mutagen An agent or process capable of inducing **mutations** (e.g. **irradiation, alkylating agents**).

mutagenesis Induction of heritable change(s) in

the genetic constitution of a cell through alterations to its **DNA**.

mutant An organism or an allele bearing a **mutation**. Usually applied when a characteristic change in **phenotype** can be recognized.

mutation Any change in the genome with respect to a defined **wild type**. Can occur at the level of **ploidy, karyotype**, or **nucleotide sequence**. Most of the latter mutations are silent (i.e. cannot be associated with any change in **phenotype**), either because the **DNA** sequence affected is in the non-coding part of the genome, or because the specific change does not alter the function of a coding sequence. *See*: **back mutation, single nucleotide polymorphism.**

mutation pressure A constant mutation rate that adds mutant genes to a population; repeated occurrences of mutations in a population.

mutualism *See*: **symbiosis.**

mycelium (pl.: mycelia) Threadlike filament making up the vegetative portion of **thallus** fungi.

mycoprotein Fungal **protein.**

mycorrhiza Fungi that form an association with, or have a symbiotic relationship with roots of more developed plants.

mycotoxin Toxic substance of fungal origin, e.g. **aflatoxin.**

myeloma A plasma cell cancer.

myo inositol *See*: **inositol.**

N

naked bud A bud not protected by bud scales.

narrow-host-range plasmid A **plasmid** that can replicate in one, or at most a few, different bacterial species.

narrow-sense heritability The proportion of the phenotypic **variance** that is due to **variation** in **breeding** values; the proportion of the phenotypic variance that is due to **additive genetic variance.**

native protein The naturally occurring form of a **protein**.

natural selection The differential survival and **reproduction** of organisms because of differences in characteristics that affect their ability to utilize environmental resources.

necrosis Death of **tissue** evidenced by discolouration, dehydration and loss of organization.

negative autogenous regulation Inhibition of the expression of a gene or set of coordinately regulated genes by the product of the gene or the product of one of the genes. *Synonym*: **negative self-regulation**.

negative control system A mechanism by which a regulatory **protein** is required to turn off **gene** expression.

negative selection Selection against individuals possessing a certain character. *Opposite*: positive selection.

negative self-regulation *See*: **negative autogenous regulation**.

nematode Slender, unsegmented worms, often parasitic. Also known as eelworm, especially when phytoparasitic.

neo-formation *See*: **organogenesis**.

neomycin phosphotransferase II (Abbreviation: npt-II). An enzyme which detoxifies the **antibiotic** neomycin, used as a **marker gene** to select for successfully transformed cells in plant **transgenesis**. *See*: ***neo***^{r}.

neoplasm Localized **cell** multiplication, forming a tumour. Generally the result of genetic **transformation**. Neoplasmic cells differ in structure and function from the original cell type.

neo^r Neomycin-resistance gene. *See*: **antibiotic resistance marker gene, neomycin phosphotransferase II, selectable marker**.

neoteny The retention of juvenile body characters in the adult state, or the occurrence of adult characters in the juvenile state.

net photosynthesis Photosynthetic activity minus respiratory activity, measured by the net absorption of carbon dioxide.

neutral mutation A mutation that changes the **nucleotide** sequence of a gene, but has no observable effect on the **fitness** of the organism.

neutral theory The theory that much of **evolution** has been primarily due to random **drift** of neutral mutations.

neutrophil A type of **leukocyte** involved in the early inflammatory response.

NFT Abbreviation for **nutrient film technique**.

nick To break (*or* a break in) a **phosphodiester bond** in one of the strands of a **double-stranded DNA** molecule.

nick translation A procedure for **labelling DNA** by treating a fragment with **DNAse** to produce single-stranded nicks followed by **excision** of a **nucleotide** and repair of the gaps with radiolabelled nucleotide.

nicked circle During the extraction of **plasmid DNA** from a bacterial cell, one strand of the **DNA** often becomes nicked. This relaxes the torsional strain which normally ensures a supercoiled structure. *Synonym*: **relaxed circle**.

***nif* gene cluster** Group of bacterial genes responsible for the biological **fixation** of atmospheric nitrogen.

nitrate The form of nitrogen that can be used directly by plants; a major component of inorganic fertilizers.

nitrification The natural process in which nitrogen in plant and animal wastes is oxidized, first to nitrites and then to nitrates, through the action of soil-borne microbes.

nitrocellulose A derivative of **cellulose**, which has the property of binding to many biological **macromolecules**, in particular **DNA**, **RNA** and **protein**. Filters made from nitrocellulose are commonly used in **Southern** and **northern blott**ing experiments. *Synonym*: **cellulose nitrate**

nitrogen assimilation The incorporation of nitrogen into the cells of living organisms.

nitrogen fixation The conversion of atmospheric nitrogen gas to oxidized forms that can be assimilated by plants, particularly by blue-green algae and some genera of bacteria (e.g. *Rhizobium* spp.; *Azotobacter* spp.). An important source of nitrogen in unfertilized soils. *See*: **nif gene cluster**.

nitrogenous base The purines (adenine and guanine) and pyrimidines (**cytosine**, and **thymine** or **uracil**) that are present in **DNA** and **RNA**.

NO Abbreviation for **nucleolar organizer**.

***nod* box** A **DNA** sequence that controls the transcriptional regulation of ***Rhizobium* nodulation** genes.

nodal culture The culture of a lateral **bud** and a section of adjacent **stem** tissue.

node Slightly swollen structure on the stem, where leaves and buds arise and where branches originate. Stems have nodes but roots do not.

nodular Term commonly used to describe the pebbly (rough) texture of a callus.

nodulation The formation of **nodules** on the roots of legumes following colonization by symbiotic nitrogen-fixing bacteria.

nodule Swollen globular structures formed on the roots of legumes, containing the nitrogen-fixing bacteria.

non repetitive DNA/RNA A nucleotide sequence which does not include a significant proportion of repetitive sequences of **nucleotides**.

non-additive genetic variation The proportion of the total genetic variation in a **population** that does not respond to simple **mass selection** and that causes specific pairwise crosses to depart from performance values predicted by the breeding values of the parents.

non-autonomous A term referring to biological units that cannot function by themselves; such units require the assistance of another unit, or "helper". *Opposite*: **autonomous**.

non-coding strand *See*: **antisense DNA**.

non-disjunction Failure of separation of **homologous** chromosomes or **chromatids** in **mitosis** or **meiosis**, resulting in too many chromosomes in some daughter cells and too few in others.

non-histone chromosomal protein In chromosomes, all of the **proteins** except the **histones**.

nonsense mutation A mutation which converts an amino-acid-specifying **codon** into a **stop codon**, e.g. a single base change from UAU to UAG generates the premature termination of the **polypeptide** chain at the position where a tyrosine was incorporated in the **wild type**.

non-target organism An organism which is affected by a treatment (e.g. **pesticide** application) for which it was not the intended recipient.

non-template strand The non-transcribed strand of **DNA**. *Synonyms*: sense strand, coding strand.

non-virulent agent *See*: **attenuated vaccine**.

NOR Abbreviation for **nucleolar organizer region.**

northern blot A technique analogous to Southern blotting, but involving the transfer of **RNA** from an **agarose gel** to a membrane, prior to probing.

npt-II Abbreviation for **neomycin phosphotransferase II.**

nucellar embryo An embryo which has developed vegetatively from somatic tissue surrounding the **embryo sac**, rather than by **fertilization** of the **egg** cell.

nucellus Tissue forming the major part of the young **ovule** in which the **embryo sac** develops; Also known as megasporangium.

nuclear transfer A technology by which novel animals are generated by **cloning** a single **diploid** somatic cell. It involves inserting a single diploid cell from a culture of cells into an enucleated ovum. The resultant diploid ovum develops into an **embryo** that is placed in a recipient female, which gives birth to the cloned animal in the normal manner. Note that the term is somewhat of a misnomer, since it is a whole cell that is transferred, not just the nucleus.

nuclease A class of largely bacterial enzymes that degrade **DNA** or **RNA** molecules by catalysing the cleavage of the **phosphodiester bonds** that link adjacent **nucleotides**. For **deoxyribonuclease** (DNAse) the substrate is DNA, for ribonuclease (**RNAse**) the substrate is RNA, and for S1 nuclease, the substrate is **single-stranded DNA** or RNA. Endonucleases **cleave** at internal sites in the substrate molecule, while exonucleases progressively cleave from the end of the substrate molecule. Nucleases have varying degrees of base-sequence specificity, the most specific being the **restriction endonucleases.**

nucleic acid A **macromolecule** consisting of polymerized **nucleotides**. Two forms are found, **DNA** and **RNA**. Nucleic acids may be linear or circularized, and single- or double-stranded. *See*: **helix.**

nucleic acid probe *See*: **DNA probe.**

nuclein The term used by Friedrich Miescher to describe the nuclear material he discovered in 1869, which today is known as **DNA**.

nucleo-cytoplasmic ratio In a cell, the ratio of nuclear to cytoplasmic volume. This ratio is high in meristematic cells and low in differentiated cells.

nucleolar organizer (Abbreviation: NO). *See*: **nucleolar organizer region.**

nucleolar organizer region (Abbreviation: NOR). A chromosomal segment containing a large array of genes that encode ribosomal **RNA**; located at the secondary constriction of specific chromosomes.

nucleolus An **RNA**-rich nuclear **organelle** in the **nucleus** of eukaryotic cells, produced by a nucleolar organizer. It represents the storage place for **ribosomes** and ribosome precursors. The nucleolus consists primarily of ribosomal precursor **RNA**, ribosomal RNA, their associated **proteins**, and some, perhaps all, of the enzymatic equipment (RNA polymerase, RNA methylase, RNA cleavage enzymes) required for synthesis, conversion and assembly of **ribosomes**. Subsequently the ribosomes are transported to the cytoplasm.

nucleoplasm The non-staining or slightly chromophilic, liquid or semi-liquid, ground substance of the **interphase nucleus** and which fills the nuclear space around the chromosomes and the nucleoli. Little is known of the chemical composition of this ground substance, which is not easily defined. Sometimes called "karyoplasm" when it is gel-like, and "karyolymph" when it is a colloidal fluid.

nucleoprotein Conjugated **protein** composed of **nucleic acid** and **protein**; the material of which the chromosomes are made.

nucleoside A **base** (**purine** or **pyrimidine**) that is covalently linked to a 5-carbon (pentose) sugar. When the sugar is **ribose**, the nucleoside is a ribonucleoside; when it is 2-deoxy**ribose**, the nucleoside is a deoxyribonucleoside. *See*: **nucleotide**.

nucleoside analogue A synthetic molecule that resembles a naturally occurring **nucleoside**.

nucleosome Spherical sub-units of eukaryotic **chromatin** that are composed of a core particle consisting of an octamer of **histones** (two molecules each of histones H_{2a}, H_{2b}, H_3 and H_4) and 146 bp of **DNA**.

nucleotide A **nucleoside** with one or more phosphate groups linked at the 3'- or 5'-hydroxyl of a pentose sugar. When the sugar is **ribose**, the nucleotide is a ribonucleotide; when it is 2-**deoxyribose**, the nucleotide is a deoxyribonucleotide. **RNA** and **DNA** are polymers of, respectively, **ribonucleoside** 5'-monophosphates and deoxyribonucleoside 5'-monophosphates. Nucleotides containing the **bases adenine**, **guanine** and **cytosine** (A, G, C) occur in both DNA and RNA; **thymine** (T) occurs only in DNA, and **uracil** (U) only in RNA. Ribonucleoside mono-, di-, and triphosphates for which a specific base is not assigned are abbreviated NMP, NDP, and NTP, while deoxyribonucleoside mono-, di-, and tri-phosphates are abbreviated dNMP, dNDP, and dNTP. Otherwise, the "N" is replaced by the base letter abbreviation. *See*: **base pair**.

nucleotide sequence *See*: **sequence**.

nucleus A dense protoplasmic membrane-bound region of a eukaryotic **cell** that contains the chromosomes, separated from the **cytoplasm** by a membrane; present in all eukaryotic cells except mature sieve-tube elements and red blood cells.

null allele An allele which produces no functional gene product.

null mutation *See*: **amorph**.

nullisomic (Adj.) *See:* **nullisomy**.

nullisomy An otherwise diploid cell or organism lacking both members of a **homologous chromosome** pair.

nurse culture Culturing cells from a **suspension culture** on a raft of filter paper above a callus **tissue** piece (nurse tissue). The filter paper serves to prevent tissue union but allows the flow of essential substances from the nurse to the isolated cells.

nutriceutical A conventional food product that has been modified (potentially by genetic engineering) to have improved nutritional characteristics and/or pharmaceutical properties.

nutrient cycle The passage of a nutrient or element through an ecosystem, including its assimilation and release by various organisms and its transformation into various organic or inorganic chemical forms.

nutrient deficiency Absence or insufficiency of an essential factor for normal growth and development.

nutrient film technique (Abbreviation: NFT). A hydroponic technique used to grow plants. NFT delivers a thin film of water or nutrient solution either continuously or through on-off cycles.

nutrient gradient A diffusion gradient of nutrients and gases that develops in tissues where only a portion of the **tissue** is in contact with the medium. Gradients are less likely to form in liquid media than in callus cultures.

nutrient medium (pl.: nutrient media) A solid, semi-solid or liquid **formulation** for *in vitro* cell or tissue growth.

O

ochre stop codon *See*: **stop codon**.

octoploid. An organism, or a tissue whose cells contain eight **haploid** sets of chromosomes.

oestrogen The generic term for a group of female **sex hormones** which control the development of sexual characteristics and control oestrus. *Alternative spelling*: estrogen.

oestrous (adj.) *See*: **oestrus**.

oestrous cycle The cycle of reproductive activity shown by most sexually mature non-pregnant female mammals. *Alternative spelling*: estrous.

oestrus In female mammals, the period of sexual excitement and acceptance of the male. *Synonyms*: rut, heat. *Alternative spelling*: estrus (adj.: estrous).

offset Young plant produced at the base of a mature plant.

offshoot Short, usually horizontal, **stem** produced near the **crown** of a plant.

offspring New individuals resulting from sexual or asexual reproduction. *Synonym*: progeny.

Okazaki fragment Replicated **DNA** fragments

formed on the **lagging strand** in **DNA** synthesis from a double-stranded **template**. The fragments are subsequently joined together by **DNA ligase**. *See*: **primosome**.

OLA Abbreviation for **oligonucleotide ligation assay**.

oligomer A molecule formed by the covalent joining of a small (undefined) number of **monomers**. *See:* **polymer**.

oligonucleotide A **nucleotide** oligomer. Often synthesized for use as **primers** for *in vitro* **DNA** synthesis. *See*: **polymerase chain reaction**.

oligonucleotide ligation assay (Abbreviation: OLA). A diagnostic technique for determining the presence or absence of a **single nucleotide polymorphism** within a **target DNA** sequence, often indicating whether the gene is wild type (normal) or **mutant** (usually defective).

oligonucleotide-directed mutagenesis *See*: **site-specific mutagenesis**.

oligonucleotide-directed site-specific mutagenesis *See*: **site-specific mutagenesis**.

oligosaccharide Carbohydrate consisting of several linked **monosaccharide** units.

oncogene A gene that causes cells to grow in an uncontrolled (i.e. tumourous) manner. Oncogenes are **mutant** forms of normal functional genes (called proto-oncogenes) that have a role in regulating **cell** proliferation. *See*: **cellular oncogene; dominant oncogene; immortalizing oncogene; recessive oncogene, p53 gene**.

oncogenesis The progression of cytological, genetic and cellular changes that culminate in the development of a tumour.

onco-mouse A mouse that has been genetically modified to incorporate an oncogene, which acts as an animal **model** for studies of human cancer.

ontogeny Developmental life history of an organism.

oocyte The **egg** mother cell; it undergoes two meiotic divisions (**oogenesis**) to form the egg cell. The **primary** oocyte refers to the state before completion of the first meiotic division; the **secondary** oocyte after completion of the first meiotic division.

oogenesis The formation and growth of the **egg** in an animal **ovary**.

oogonium 1. A **germ cell** of the female animal, that gives rise to oocytes by mitotic division. 2. The female sex organ of algae and fungi.

oosphere The non-motile female **gamete** in plants and some algae.

oospore A spore developing from a **zygote** of certain algae and fungi, following the fusion of **heterogametes**.

opal stop codon *See*: **stop codon**.

open continuous culture A continuous culture system, in which inflow of fresh medium is balanced by **outflow** of a corresponding volume of **spent medium** plus cells. In the steady state, the rate of cell **wash-out** equals the rate of formation of new cells in the system. *See*: **continuous culture; batch culture; closed continuous culture**.

open pollination Pollination by wind, insects or other natural mechanisms.

open reading frame (Abbreviation: ORF). A sequence of nucleotides in a **DNA** molecule that has the potential to **encode** a **peptide** or **protein**: comprisesa start **triplet** (ATG), followed by a series of triplets (each of which encodes an **amino acid**), and ending with a **stop codon** (TAA, TAG or TGA). The term is generally applied to sequences of **DNA** fragments, for which no function has yet been determined. The number of ORFs provides an estimate of the number of genes transcribed from the **DNA** sequence.

operational definition An operation or procedure that can be carried out to define or delimit something.

operator The region of **DNA** that is **upstream** from a **gene** or genes and to which one or more regulatory **proteins** (repressor or activator) bind to control the expression of the gene(s).

operon A functionally integrated genetic unit for the control of **gene expression** in bacteria. It consists of one or more genes that **encode** one or more **polypeptide**(s) and the adjacent site (**promoter** and **operator**) that controls their expression by regulating the **transcription** of the structural genes.

opine The condensation product of an **amino acid** with either a keto-acid or a sugar, produced by the plant host as a result of *Agrobacterium* infection, and used exclusively by the *Agrobacterium* as a carbon source for growth and **reproduction** within the plant.

OPU Abbreviation for **ovum pickup**.

ORF Abbreviation for **open reading frame**.

organ A tissue or group of **tissues** that constitute a morphologically and functionally distinct part of an organism.

organ culture The **aseptic** culture of complete living organs of animals and plants outside the body in a suitable **culture medium**. Animal organs must be small enough to allow the nutrients in the culture medium to penetrate all the cells.

organellar gene Genes located on organelles outside the nucleus.

organelle A membrane-bounded specialized region within a cell, such as the **mitochondrion** or dictyosome, that carries out a specialized function in the life of a cell.

organic complex A chemically undefined compound added to nutrient **media** to stimulate growth, e.g. coconut milk; malt extract; **yeast** extract; **casein** hydrolysate.

organic evolution *See:* **evolution**.

organism An individual living system, such as animal, plant or microorganism, that is capable of reproduction, growth and maintenance.

organized growth The development under **tissue culture** conditions of organized **explants** (meristem tips or shoot tips, floral buds or organ primordia). *Opposite*: **unorganized growth**.

organized tissue Composed of normally differentiated cells.

organogenesis The initiation of **adventitious** or *de novo* shoots or roots from **callus, meristem**

or suspension cultures. *See*: **micropropagation**; **regeneration**.

organoid An organ-like structure produced in culture.

organoleptic Having an effect on one of the organs of sense, such as taste or smell.

origin of replication The **nucleotide** position on a **DNA** sequence from which **DNA** synthesis (replication) is initiated.

orphan gene A gene or **DNA sequence** whose function is not known.

orphan receptor A **receptor** for which a cellular function or **ligand** has yet to be identified.

ortet The plant from which a **clone** is obtained. *Synonym*: **donor plant**.

orthologous Homologous genes/gene products that have evolved divergently *between* species; many rice genes have orthologues in other cereal genomes, because of the common ancestry of cereal species. *See*: **paralogous**.

osmosis Diffusion of water across a **semi-permeable** membrane from zones of low solute concentration to zones of higher solute concentration.

osmotic potential Change in the energy state of solvent brought about by dissolving a substance in the solvent - water in the biological sciences. The potential of aqueous solutions is always negative compared to pure water. Solvent flows from higher to lower osmotic potential solutions by **diffusion** or **osmosis**.

osmoticum A chemical agent (e.g. **polyethylène glycol, mannitol**, glucose or sucrose) employed to maintain the **osmotic potential** of a nutrient medium equivalent to that of the cultured cells, i.e. the medium and the cells are **isotonic**. Because of this osmotic equilibrium, cells are not damaged *in vitro*.

outbreeding A mating system characterized by the inter-breeding of genetically unrelated or dissimilar individuals. Since **genetic diversity** tends to be enhanced, and vigour or **fitness** of individuals increased by this process, it is often used to counter the detrimental effects of continuous inbreeding. *Synonym*: **exogamy**.

outflow The volume of growing cells that is

removed from a **bioreactor** during a **continuous fermentation** process.

ovary 1. Enlarged basal portion of the **pistil** of a plant flower that contains the **ovules**. 2. The reproduction **organ** in female animals in which eggs are produced.

overdominance Where **heterozygote** performance is superior to that of either parental **genotype**.

overhang *See*: **extension**.

overlapping reading frame Start **codons** in different **reading frames** generate different **polypeptides** from the same **DNA** sequence.

ovulation The release of mammalian egg(s) from the ovary.

ovule The part of the **seed** plant's reproductive organs that comprises the **nucellus**, the **embryo sac** and the integuments.

ovum (pl.: ova) Egg.

ovum pickup (Abbreviation: OPU). The non-surgical collection of ova from a female.

oxidative phosphorylation The enzymatic addition of a phosphate to **ADP** to make **ATP**, coupled to electron transport from a **substrate** to molecular oxygen. A critical reaction for the generation of cellular energy.

oxygen-electrode-based sensor Sensor in which an oxygen electrode, which measures the amount of oxygen in a solution, is coated with a biological material such as an **enzyme** which generates or absorbs oxygen when the appropriate **substrate** is present. When the biological reaction takes place, the amount of oxygen close to the electrode changes and the signal from the electrode changes, thus signalling the concentration of substrate.

P

p Denotes the shorter of the two **chromosome** arms, e.g. human 14p is the short arm of human chromosome 14.

P element A *Drosophila* **transposon**.

$\mathbf{P_1}$, $\mathbf{P_2}$ Generational symbols for the two parents of a given individual.

p53 gene A human tumour-suppressor **transcription** factor **gene**, damage or **mutation** to which is believed to be responsible for up to 60% of all human cancer tumours. If, in spite of the presence of p53 **protein**, a cell begins to divide uncontrollably following damage to its **DNA**, the p53 gene acts to prevent tumours by triggering **apoptosis**.

pachynema A mid-prophase stage in the first meiotic division, between **zygonema** and **diplonema**. Chromosomes appear as long, paired threads. Occasionally, all four **chromatids** can be recognized.

pachytene (adj.) *See:* **pachynema**.

packaging cell line A cell line designed to produce viral particles that do not contain nucleic acid. After **transfection** of these cells with a full-size viral genome, fully infective viral particles are assembled and released.

packed cell volume (Abbreviation: PCV). The proportion of a **cell culture** volume that is occupied by cells. Cell volume is determined by sedimenting using low speed centrifugation.

PAGE Abbreviation for **polyacrylamide gel electrophoresis.**

pairing The pairing of **homologous** chromosomes during the **prophase** of the first meiotic division. Pairing is the first prerequisite before **crossing over** and **recombination** can occur. *Synonym*: **synapsis.**

pair-rule gene A gene that influences the formation of body segments in *Drosophila.*

palaeontology The study of the fossil record of past geological periods and of the phylogenetic relationships between extinct and contemporary plant and animal species.

palindrome A segment of double-stranded **DNA,** in which the order of bases, read 5'?3' in one strand, is the same as that in the **complementary** antiparallel strand, also read 5'?3'. If the sequence is written in the normal convention, on two lines with paired bases shown one above the other, the base order on one strand runs in the opposite direction to that on the complementary strand. They are often found at the ends of **transposable elements,** and recognition sites for type II **restriction endonucleases** are also palindromes. *Synonym*: **inverted repeat.**

palisade parenchyma Elongated cells found just beneath the upper **epidermis** of leaves, typically rich in **chloroplasts.**

pAMP Ampicillin-resistant **plasmid.**

panicle An inflorescence, the main axis of which is branched; the branches bear loose racemose flower clusters. Rice is a prominent crop plant with a panicle inflorescence.

panicle culture Aseptic culture of immature panicle **explants** to induce **microspore** germination and development.

panmictic population A population in which mating occurs at random.

panmixis Random mating in a population.

paper raft technique *See*: **nurse culture**.

PAR Abbreviation for **photosynthetically active radiation**.

***par* gene** One of a class of genes required for faithful **plasmid segregation** at **cell** division. Initially, par loci were identified on plasmids, but have also been found on bacterial chromosomes.

paracentric inversion A chromosomal rearrangement in which a segment of chromosome, excluding the **centromere**, is rotated.

paraffin [wax] A translucent, white, solid hydrocarbon with a low melting point. One use is as an embedding medium to support **tissue** for sectioning for light microscopy observation.

Parafilmä A stretchable film based on paraffin wax; used to seal tubes and Petri dishes. Parafilmä is a proprietary name which is incorrectly applied colloquially to similar products.

parahormone A substance with hormone-like properties that is not a secretory product (e.g. ethylene; carbon dioxide).

parallel evolution The development of different organisms along similar evolutionary paths due to similar **selection** pressures acting on them.

paralogous Homologous genes/gene products that have duplicated and evolved divergently *within* a species. e.g. beta- and gamma-globulin genes.

parameter A value or measurement that varies with circumstances, and is used as a reference to quantify a situation or a process.

parasexual cycle A sexual cycle involving changes in **chromosome** number but differing in time and place from the usual sexual cycle; occurring in those fungi in which the normal cycle is suppressed or apparently absent.

parasexual hybridization *See*: **somatic hybridization**.

parasite An organism deriving its food from the living body of another organism.

parasitism The close association of two or more dissimilar organisms, where the association is harmful to the **host**, but beneficial to the **parasite**.

parasporal crystal Tightly packaged insect **protoxin** molecules that are produced by strains of ***Bacillus thuringiensis*** during the formation of resting spores.

paratope Synonym for **antibody binding site**.

parenchyma 1. A plant tissue consisting of spherical, undifferentiated cells, frequently with air spaces between them. 2. Loose connective **tissue** formed by large cells.

parenchymatous (adj.) *See*: **parenchyma**.

parthenocarpy The development of fruit without fertilization.

parthenogenesis Production of an **embryo** from an unfertilized egg.

partial digest Incomplete reaction of a **restriction enzyme** with **DNA**, such that only a proportion of the target sites are cleaved. Partial digests are often performed to give an overlapping collection of **DNA** fragments for use in the construction of a gene bank. *Synonym*: **incomplete digest**. *Opposite*: **complete digest**.

particle radiation High energy nuclear emissions, used as physical agents of **mutagenesis**. Three major types typically used: alpha (a) (positively charged), beta (b) (negatively charged), or neutrons (uncharged).

parts per million (Abbreviation: ppm). Unit of concentration: 1ppm = 1mg dissolved matter per litre of solution.

parturition The process of giving birth.

passage number The number of times cells in culture have been sub-cultured.

passage The transfer or transplantation of cells from one **culture medium** to another.

passage time Time interval between successive sub-cultures.

passive immunity 1. Natural acquisition of antibodies by the **foetus** or neonate (newborn) from the mother. 2. The artificial introduction of specific antibodies by the injection of **serum** from an immune animal. In both cases, temporary protection is conferred on the recipient. *See*: **immunization**, **immunoprophylaxis**.

***pat* gene** A gene obtained from *Streptomyces* sp. encoding resistance to glufosinate-ammonium containing herbicides. (These inhibit plant synthesis of glutamine). Used widely as a means of transgenically inducing **herbicide resistance** in crop plants. *Synonym*: ***bar* gene**.

patent A legal permission to hold exclusive right - for a defined period of time - to manufacture, use or sell an invention.

paternal Pertaining to the father.

pathogen A disease-causing organism (generally microbial: bacteria, fungi, viruses; but can extend to other organisms: e.g. nematodes etc.). *Synonym*: infectious agent. *See*: **latent agent**.

pathogenesis related protein (Abbreviation: PR **protein**). One of a group of **proteins** that are characteristically highly expressed as part of a plant's response to **pathogen** attack. Many of these **proteins** are similarly expressed following **infection** with a broad range of pathogens, indicating their role in the **hypersensitive response**.

pathogen-free Uncontaminated with pathogen.

pathotoxin Substance secreted by certain **pathogens**, in order to attack the **host** tissue. Some pathotoxins are also toxic to non-hosts, especially animals or humans.

pathovar Strain of a plant-attacking bacterium or fungus that can be differentiated from others by their interaction with specific **host cultivars**.

PBR Abbreviation for **plant breeders' rights**.

pBR322 One of the first **plasmids** used for cloning **DNA** in *E. coli*.

PCR Abbreviation for **polymerase chain reaction**.

PCR-RFLP Alternative term for **cleaved amplified polymorphic sequence**.

PCV Abbreviation for **packed cell volume**.

pectin A group of naturally occurring complex **polysaccharides**, containing galacturonic acid, found in plant **cell walls**, where their function is to cement cells together. Used as a thickening agent in solid **culture media** and as a food additive.

pectinase Enzyme catalysing the **hydrolysis** of **pectin**. Used in conjunction with **cellulase** to solubilize plant **cell walls**.

pedicel Stalk or **stem** of the individual flowers of an inflorescence.

pedigree The ancestry of an individual.

peduncle Stalk or **stem** of a flower that is born singly; the main stem of an inflorescence.

PEG Abbreviation for **polyethylene glycol**.

penetrance The proportion of individuals in a **population** that **express** the **phenotype** expected from their **genotype** with respect to a specific gene. Measures the extent to which phenotype predicts genotype.

peptidase An enzyme that catalyzes the **hydrolysis** of a **peptide** bond.

peptide A sequence of **amino acids** linked by **peptide bonds**; a breakdown or build-up unit in **protein** metabolism. Typically used to describe low molecular weight species. *See*: **polypeptide**.

peptide bond The chemical bond holding **amino acid** residues together in peptides and **proteins**. The (CO-NH) bond is formed by the condensation, with loss of a water molecule, between the carboxyl (-COOH) group of one amino acid and the amino ($-NH_2$) of the next amino acid.

peptide expression library A collection of **peptide** molecules, produced by **recombinant** cells, in which the **amino acid** sequences are varied.

peptide nucleic acid (Abbreviation: PNA). A synthetic **oligonucleotide** analogue, in which the sugar backbone is replaced by a **peptide** chain, upon which the **nucleoside** residues are strung. **Probes** made from PNA appear to have greater **specificity** than those made from **DNA**.

peptide vaccine A short chain of amino acids that can induce antibodies against a specific infectious agent.

peptidyl transferase An enzyme bound tightly

to the large sub-unit of the ribosome, that catalyses the formation of **peptide** bonds between **amino acids** during translation.

peptidyl-tRNA binding site (Abbreviation: P-site). The site on a **ribosome** that hosts the **tRNA** to which the next **amino acid** for the growing **polypeptide** chain is attached.

perennial A plant that flowers continuously for several years.

pericentric inversion A chromosomal rearrangement in which a segment of chromosome, including the **centromere**, is rotated.

periclinal chimera 1. Genotypically or cytoplasmically different tissues arranged in concentric layers. 2. A **chimera** in which one or more layers of tissue derived from one **graft** member enclose the central **tissue** derived from the other member of the graft.

periclinal The orientation of **cell wall** or plane of **cell division** parallel to the reference surface.

pericycle Region of the plant bounded externally by the **endodermis** and internally by the **phloem**. Most roots originate from the pericycle.

periplasm The space between the **cell** (cytoplasmic) membrane of a bacterium or fungus and the outer membrane or cell wall. *Synonym*: periplasmic space.

permanent wilting point (Abbreviation: PWP). The moisture content of a soil below which plants wilt to such an extent that they fail to recover even when fully watered.

permeable Referring to a membrane, cell or cell system through which small molecules can diffuse.

persistence Ability of an organism to remain in a particular setting for a period of time after it is introduced.

persistent Chemicals with a long inactivation or degradation time, such as some pesticides. Persistent substances can become dangerously concentrated in the tissues of organisms at the top end of a food chain.

PERV Abbreviation for **porcine endogenous retrovirus**.

pesticide A toxic chemical product that kills

harmful organisms (e.g. insecticides, fungicide, weedicides, rodenticides).

petal One of the parts of the flower that make up the corolla.

petiole Stalk of leaf. *See*: **pedicel; peduncle.**

Petri dish Flat round glass or plastic dish with a matching lid, used for small-scale culturing of organisms, germinating seeds etc. Also referred to as plates, hence to **plate** a culture.

PFGE Abbreviation for **pulsed-field gel electrophoresis.**

PG Abbreviation for **polygalacturonase.**

pH Logarithmic measure of acidity/alkalinity of a solution. A pH of 7 is neutral (e.g. pure water), whereas below 7 is acid and above 7 is alkaline.

phage Abbreviation for **bacteriophage.**

phagemids Cloning vectors that contain components derived from both **phage** and **plasmid DNA.**

phagocytes Immune system cells that ingest and destroy viruses, bacteria, fungi and other foreign substances or cells.

phagocytosis The process by which foreign particles invading the body are engulfed and broken down by **phagocytes.**

pharmaceutical agent *See*: **therapeutic agent.**

pharmacokinetics The quantitative measurement of how drugs move around the body, and the processes which control their absorption, distribution, metabolism, and excretion.

phase change The developmental change from one **maturation** state to another.

phase state The **coupling** or **repulsion** of two linked genes.

PHB Abbreviation for **polyhydroxybutyrate.**

pH-electrode-based sensor Sensor in which a standard **pH** electrode is coated with a biological material. Many biological processes raise or lower pH, and the changes can be detected by the pH electrode.

phenocopy An environmentally induced, non hereditary **variation** in an organism, resembling a genetically determined trait.

phenolic oxidation Common aspect of the wound response in plants. Phenolic oxidation is often indicated by blackening of **tissue** and it may be a precursor to growth inhibition or, in severe cases, to tissue **necrosis** and death.

phenolics Compounds with hydroxyl group(s) attached to the benzene ring, forming esters, ethers and salts. Phenolic substances produced from newly explanted tissues are liable to oxidise, and as a result form coloured compounds visible in nutrient media.

phenotype The visible appearance of an individual (with respect to one or more traits) which reflects the reaction of a given **genotype** with a given environment.

pheromone A hormone-like substance that is secreted by an organism into the environment as a specific signal to another organism, usually of the same species.

phloem Specialized **vascular plant** tissue for the transport of assimilates (generally sugars) from the point of synthesis (in the leaf) to other parts of the plant. It consists of sieve tubes, companion cells, phloem parenchyma and fibres.

phosphatase An class of enzymes that catalyze the **hydrolysis** of esters of phosphoric acid, removing a phosphate group from an organic compound.

phosphodiester (phospho-diester) bond A bond in which a phosphate group joins adjacent carbons through ester linkages. A condensation reaction between adjacent nucleotides results in a phosphodiester bond between 3' and 5' carbons in **DNA** and **RNA**.

phospholipase A2 An enzyme which degrades type A2 **phospholipids**.

phospholipid A class of **lipid** molecules in which glycerol is linked to a phosphate group and two fatty acyl groups. Contains both polar and non-polar regions. A major component of biological membranes. *See*: **inositol lipid**.

phosphorolysis The cleavage of a bond by orthophosphate; analogous to **hydrolysis** referring to cleavage by water.

phosphorylation The addition of a phosphate group to a compound.

photoautotroph *See*: **autotroph, heterotroph.**

photo-bioreactor Bioreactor dependent on sunlight, which is taken up by its content of plant material, usually algae.

photoheterotroph *See*: **heterotroph.**

photoperiod The length of daylight or period of daily illumination provided for growth.

photoperiodism The photoperiod required by a plant to switch from the vegetative to the reproductive stage.

photophosphorylation The formation of **ATP** from **ADP** and inorganic phosphate using light energy harvested by photosynthesis.

photoreactivation A light dependent **DNA repair** process.

photosynthate The carbohydrates and other compounds produced in photosynthesis.

photosynthesis A chemical process by which green plants synthesize organic compounds from carbon dioxide and water in the presence of sunlight.

photosynthetic Able to use sunlight energy to convert atmospheric carbon dioxide into organic compounds. Nearly all plants, most algae and some bacteria are photosynthetic.

photosynthetic efficiency Efficiency of converting light energy into organic compounds.

photosynthetic photon flux (Abbreviation: PPF). A measure of the intensity of light utilized by plants for **photosynthetic** activity.

photosynthetically active radiation (Abbreviation: PAR). The part of the radiant energy that is capturable by natural **photosynthetic** systems (approximately equivalent to the natural light spectrum of wavelengths 400-700nm).

phototropism The tendency of plants to direct shoot growth towards the source of light.

phylogeny The deduced evolutionary history of related organisms.

physical map An indication of the separation, in bp, between pairs of linked loci. *See*: **mapping.**

phyto- (Prefix) To do with plants.

phytochemical Molecules characteristically found in plants.

phytochrome A pigment, found in the **cyto-**

plasm of green plants, which can exist in two forms Pr (biologically inactive) and Pfr (biologically active). Pfr is converted into Pr by exposure to light of wavelength 730 nm. Involved in the timing of many plant processes, e.g. dormancy, leaf formation, flowering and germination.

phytohormone A substance that stimulates growth or other processes in plants. Major species are **auxins, abscisic acid, cytokinins, gibberellins** and **ethylene**.

phytokinin *See*: **cytokinin**.

phytoparasite Parasite on plants.

phytoparasitic (adj.) *See:* **phytoparasite**.

phytopathogen A plant **pathogen**.

phytoremediation The use of plants actively to remove contaminants or pollutants from either soils (e.g. polluted fields) or water resources (e.g. polluted lakes). An example is the exploitation of the Brazil water hyacinth (*Eichhornia crassipes*) to accumulate in its tissues toxic metals such as lead, arsenic, cadmium, mercury, nickel, and copper.

phytosanitary Plant health, including quarantine.

phytostat Apparatus designed for the semi-continuous chemostatic culture of plant cells.

phytosterol One of a group of biologically active **phytochemicals** present in the seeds of certain plants. Evidence suggests that human consumption of certain phytosterols, such as â-sitosterol, can help to lower total **serum** cholesterol and low-density lipo**proteins** levels, thereby reducing the risk of coronary heart disease.

pigment Compounds that are coloured by the light they absorb. Light absorption is exploited by plants both as a means of energy capture (*see*: **photosynthesis**) and as a signalling mechanism (*see*: **phytochrome**).

pinocytosis The engulfing of a minute droplet of liquid by a living cell.

pipette Widely used device for accurate dispensing of small volumes of liquids.

pistil Central organ of the flower, typically consisting of **ovary**, **style** and stigma. Usually referred to as the female part of a perfect flower.

plant breeders' rights (Abbreviation: PBR). Legal protection of a new plant **variety** granted to the breeder or his successor in title. The effect of PBR is that prior authorization is required before the material can be used for commercial purposes.

plant cell culture *In vitro* growth of plant cells.

plant cell immobilization Entrapment of plant cells in **gel** matrices so that they are protected from physical damage. The cells are suspended in liquified droplets which are then allowed to harden. Commonly used matrices are alginates, **agar** or polyacrylamide.

plant genetic resources (Abbreviation: PGR). The reproductive or vegetative propagating material of: 1. cultivated varieties (cultivars) in current use and newly developed varieties; 2. obsolete **cultivars**; 3. primitive cultivars (landraces); 4. wild and **weed** species, near relatives of cultivated varieties; and 5. special genetic stocks (including elite and current breeder's lines and mutants).

plant growth regulator An organic compound, either natural or synthetic, and other than a nutrient, that modifies or controls one or more specific physiological processes within a plant.

plant hormone *See*: **plant growth regulator**.

plant variety protection (Abbreviation: PVP). Synonym for **plant breeders' rights**.

plant variety rights *See*: **plant breeders' rights**.

plantibody An antibody expressed transgenically in an engineered plant.

plantlet A small rooted shoot regenerated from **cell culture** following **embryogenesis** or **organogenesis**. Plantlets can normally develop into normal plants when transplanted to soil.

plaque A clear spot on an otherwise opaque **lawn** of bacteria, where cells have been lysed by **phage** infection.

plasma cells Antibody-producing white blood cells derived from B **lymphocytes**.

plasma membrane *See*: **plasmalemma**.

plasma The fluid portion of the blood in which

is suspended the white and red blood cells. Contains 8-9% solids, of which 85% is composed of the **proteins** fibrinogen, albumin, and globulin. The essential function of plasma is the maintenance of blood pressure and the transport of nutrients and waste.

plasmalemma The **lipid** bilayer and associated **proteins** and other molecules that surrounding the **protoplast**, within the cell wall. *Synonyms*: **cell membrane; plasma membrane**.

plasmid An circular self-replicating non-chromosomal **DNA** molecule found in many bacteria, capable of transfer between bacterial cells of the same species, and occasionally of different species. **Antibiotic resistance genes** are frequently located on plasmids. Plasmids are particularly important as **vectors** for genetic engineering.

plasmodesma (pl.: plasmodesmata) Fine protoplasmic thread that connects adjacent plant cells by passing through the plant cell wall. Exploited by viruses as a conduit for cell to cell movement.

plasmolysis Shrinkage of **protoplasm** caused by removal of water from a cell through **osmosis** when surrounded by a **hypertonic** solution.

plastid A general term for a number of plant **cell organelles** which carry non-nuclear **DNA**. Includes the **pigment**-carrying bodies: 1. **chloroplasts** in leaves, 2. **chromoplasts** in flowers, and 3. the starch-synthesizing amyloplasts in seeds.

plastoquinone One of a group of compounds involved in the transport of electrons as part of the process of **photosynthesis**.

plate 1. Verb: to distribute a thin film of microorganisms or plant cells onto a solid medium. 2. Noun: refers to the two segments of a **Petri dish** or similar.

platform shaker *See*: **shaker**.

plating efficiency The percentage of inoculated cells which give rise to cell colonies when seeded into **culture** vessels.

pleiotropic (adj.) *See*: **pleiotropy.**

pleiotropy The simultaneous effect of a given

gene on more than one apparently unrelated trait.

ploidy The number of complete sets of chromosomes per cell, e.g. one set: **haploid**, two sets: **diploid**, etc.

plumule The first **bud** of an **embryo**, or that portion of the young shoot above the **cotyledons.**

pluripotent *See*: **totipotent.**

plus tree *See*: **elite tree.**

PNA Abbreviation for **peptide nucleic acid.**

pneumatic reactor *See*: **airlift fermenter.**

point mutation A change in **DNA** sequence at a specific **locus**. The smallest change involves the substitution, **deletion** or insertion of a single **nucleotide**. *See*: **single nucleotide polymorphism.**

polar bodies In female animals, the products of a meiotic division that do not develop into a functional **ovum**. The first polar body comprises one of the two products of the first meiotic divisions, which may fail to divide at the second division. The second polar body comprises one of the products of the second division.

polar mutation A mutation that influences the functioning of genes that are **downstream** from the site of **mutagenesis** but are in the same **transcription** unit.

polar nuclei Two centrally located nuclei in the **embryo sac** that unite with a second **sperm** cell in a triple fusion. In some plant species (particularly the **monocotyledons**), the product of this fusion develops into the endosperm.

polar transport A directed movement within plants of compounds (usually endogenous plant growth regulators) mostly in one direction; polar transport overcomes the tendency for **diffusion** in all directions.

polarity The observed **differentiation** of an **organism, tissue** or **cell** into parts having opposed or contrasted properties or form.

pole cells A group of cells in the posterior of *Drosophila* embryos that are precursors to the adult **germ** line.

pollen culture The *in vitro* culture and germination of pollen grains to generate **haploid** plants. *See*: **anther culture; microspore culture**.

pollen grain The mature **microspore**, produced in the pollen sac of angiosperms or the microsporangium of gymnosperms. Unicellular, with variable shape and size, and an elaborately structured wall.

pollen Mature **microspores** of **seed** plants.

pollination Transfer of **pollen** from **anther** to **stigma** in the process of **fertilization** in angiosperms; transfer of pollen from male to female cone in the process of fertilization in gymnosperms.

poly-(A) polymerase Enzyme that catalyses the addition of **adenine** residues to the 3' end of **mRNA** molecules, forming the characteristic **poly-(A) tail**.

poly-(A) tail *See*: **polyadenylation**.

polyacrylamide gel electrophoresis (Abbreviation: PAGE). Ubiquitous method for separating nucleic acids and **proteins** on the basis of their molecular size. The method relies on the migration through an inert matrix (**polyacrylamide gel**) of electrically charged molecules as a result of the imposition of an electric field.

polyacrylamide gel Inert **electrophoresis** matrix, formed by the **polymerization** of acrylamide **monomer** in the presence of the cross-linker *N,N'*-methylene-*bis*-acrylamide. Gels are usually supported between two glass plates, which need to be removed for post-electrophoresis manipulations. Sometimes referred to incorrectly as acrylamide gels.

polyadenylation Post-**transcription**al addition of multiple **adenine** residues to the 3' end of eukaryotic **mRNA**. Also called poly-(A) tailing. The adenine-rich 3' terminal segment is called a poly (A) tail.

polycistronic A single **mRNA** that contains the information necessary for the production of more than one polypeptide. Particularly characteristic of prokaryotic **mRNAs**.

polyclonal antibody A **serum** sample that contains a mixture of distinct **immunoglobulin**

molecules, each recognizing a different **antigenic determinant** of a given antigen.

polycloning site *See*: **polylinker**.

polyembryony The production of more than one **embryo** from a single **egg** cell (in animals) or from a range of embryogenic cell types (in plants). These embryos are genetically identical to one another.

polyethylene glycol (Abbreviation: PEG). A **polymer** having the general formula $HOCH_2(CH_2OCH_2)_nCH_2OH$ and available in a range of molecular weights. Thus PEG 1000 is a polyethylene glycol of average molecular weight 1000. PEG 4000 and 6000 are commonly used to promote **cell** or **protoplast** fusion, and to facilitate **DNA** uptake in the **transformation** of organisms such as yeast. Also used to concentrate solutions by withdrawing water from them via **osmosis**.

polygalacturonase (Abbreviation: PG). An **enzyme** which catalyses the breakdown of pectin. A tomato engineered to contain an antisense-PG gene succeeded in delaying the onset of softening, by inhibiting the expression of PG. This allowed the fruit to be picked at a riper stage than is conventionally possible, and represented the first commercialized genetically engineered crop plant.

polygene One of a number of genes, each of small effect, which together act to determine the **phenotype** of a quantitative **trait**. The result is continuous **variation** in the trait and a seemingly non-Mendelian mode of inheritance. *See*: **quantitative trait locus, continuous variation.**

polygenic Character controlled by many genes of small effect. *See:* **polygene**.

polyhydroxybutyrate (Abbreviation: PHB). A **biopolymer,** with physical properties similar to polystyrene, originally discovered in the bacterium *Alcaligenes eutropus*. The **gene coding** for this compound has since been transformed into both other bacteria and into some crop plants in order to produce a source of renewable raw material for the plastics industry. It is rapidly degraded by soil microorganisms.

polylinker A synthetic segment of **DNA**, designed to include a number of different **restriction endonuclease** sites. When ligated to a **DNA fragment** that is to be cloned, this enables a wide choice of restriction endonucleases to be used for the **cloning** process. *Synonym*: **multiple cloning site** (MCS).

polymer A **macromolecule** synthesized by the chemical joining of many identical or similar **monomers**. For example, amino acids, monosaccharides and nucleotides give rise to **proteins**, polysaccharides and nucleic acids respectively. Water is eliminated between the monomers as they link to form chains. The individual monomer units condensed within a chain are often referred to as residues, a term which is also employed for the **bases** incorporated in polynucleotides.

polymerase An enzyme that catalyses the formation of **polymers** from **monomers**. A **DNA** polymerase synthesizes DNA from deoxynucleoside triphosphates using a **complementary** DNA strand and a primer. An **RNA** polymerase synthesizes RNA from ribonucleoside triphosphates and a complementary DNA strand.

polymerase chain reaction (Abbreviation: PCR). A widespread molecular biology procedure that allows the production of multiple copies (**amplification**) of a specific **DNA sequence**, provided that the base pair sequence of each end of the target is known. It involves multiple cycles of DNA denaturation, **primer** annealing, and strand extension, and requires a **thermostable** DNA polymerase, deoxyribonucleotides, and specific **oligonucleotides** (primers).

polymerization The chemical fusion of a number of identical or similar **monomers** to form a **polymer**. Common biological polymers are **starch** (polymerized **monosaccharides**), **DNA** (**deoxyribonucleotides**) and **proteins** (**amino acids**).

polymery The phenomenon whereby a number of genes at different loci (which may be **poly-**

genes) can act together to produce a single effect.

polymorphism 1. The occurrence of **allelic** variation at a **locus**. Polymorphism in nucleotide sequences has provided powerful diagnostic tools. *See:* **DNA diagnostics, microsatellites, restriction fragment length polymorphism.** 2. The occurrence of two or more forms in a population. *See:* **balanced polymorphism, chromosomal polymorphism.**

polynucleotide A linear **polymer** composed of covalently linked **nucleotides**. Each link is formed by a single **phosphodiester bond**. The term is used to describe **DNA** and **RNA**.

polypeptide A linear **polymer** composed of covalently linked amino acids. Each link is formed by a single **peptide** bond. Sometimes used as a synonym for **protein**, but also describes non-natural and low-molecular-weight polymers.

polyploid Organism, **tissue** or **cells** having more than two complete sets of chromosomes. Many crop plants are polyploid, including bread wheat (hexaploid, 6x), cotton and alfalfa (tetraploid, 4x), and banana (triploid, 3x).

polysaccharide A linear or branched **polymer** (e.g. starch, cellulose, etc.) composed of covalently linked **monosaccharides**, including **cellulose, pectin** and **starch**. *Synonym*: carbohydrate.

polysaccharide capsule *See*: **capsule.**

polysome A multi-ribosomal structure representing a linear array of **ribosomes** held together by **mRNA**.

polyspermy The entry of several **sperm** nuclei into the egg during fertilization, although only one actually fuses with the egg nucleus.

polytene chromosome Giant chromosomes produced by **interphase replication** without division, and consisting of many identical **chromatids** arranged side by side.

polyunsaturates Oils in which some of the carbon-carbon bonds are not fully hydrogenated - i.e. of the form -CH=CH-, rather than $-CH_2-CH_2-$.

polyvalent vaccine A **recombinant** organism

into which antigenic determinants have been cloned from a number of **pathogens**, for use as a **vaccine**.

polyvinylpyrrolidone (Abbreviation: PVP). An occasional constituent of plant **tissue culture** isolation media. PVP is of variable molecular weight and of general formula $(C_6H_9NO)_n$. Its **antioxidant** properties are used to prevent oxidative **browning** of excised plant tissues. Less frequently used as an **osmoticum** in culture media.

population A defined group of interbreeding organisms.

population density Number of cells or individuals per unit. The unit could be an area, or a volume of medium.

population genetics The branch of genetics that deals with frequencies of alleles and genotypes in **breeding** populations.

porcine endogenous retrovirus (Abbreviation: PERV). The **provirus** of a porcine retrovirus. The possibility that PERVs could be activated after **xenotransplantation** of pig organs into humans has raised concern that xenotransplantation may result in the transfer of novel infections to the human population.

position effect The influence of the location of a gene (particularly a **transgene**) on its expression and hence its effect on phenotype.

positional candidate gene A gene known to be located in the same region as a **DNA marker** that has been shown to be linked to a single-locus **trait** or to a **QTL**, and whose deduced function suggests that it could be the source of genetic variation in the **trait** in question.

positional cloning A strategy for gene cloning that relies on the identification of closely linked markers to the target trait, and then uses **chromosome walking** to identify, isolate and characterize the gene(s) responsible for the **trait**. The strategy is particularly appropriate when the biochemical basis of the target trait is unclear, thus precluding the use of a **candidate gene** approach.

positive control system A mechanism in which

a regulatory **protein**(s) is required to turn on **gene expression.**

positive selectable marker *See*: **dominant selectable marker.**

positive selection A method by which cells that carry a **DNA insert** integrated at a specific chromosomal location can be selected, since this integration confers a predictable **phenotype.**

post-replication repair A recombination-dependent mechanism for repairing damaged **DNA.**

post-translational modification The addition of specific chemical residues to a **protein** after it has been translated. Common residues are phosphate groups (phosphorylation) and sugars (glycosylation).

potentiometric *See*: **enzyme electrode.**

PPF Abbreviation for **photosynthetic photon flux.**

ppm Abbreviation for **parts per million.**

PR protein Abbreviation for **pathogenesis related protein.**

precautionary principle The approach whereby any possible risk associated with the introduction of a new technology is avoided, until a full understanding of its impact on health, environment etc. is available. Particularly applied to the release of genetically modified organisms, since unlike many technologies, these cannot be recalled if problems arise.

precocious germination Early germination of a **seed** or **embryo**, prior to the full maturation of the embryo.

pre-filter A coarse filter used to screen out large particles from a fluid or gas, before it is passed through a much finer filter.

pre-mRNA *See*: **primary transcript.**

pressure potential The pressure generated within a cell, being the net difference between the cell's **osmotic** potential and the **water potential** of the external environment.

pre-transplant A stage in **micropropagation** - the rooting and hardening process prior to transfer to soil.

preventive immunization Infection with an **antigen** to elicit an **antibody** response that will protect the organism against future infections. *Synonym*: vaccination.

Pribnow box Consensus **sequence** near the **mRNA** start-point of prokaryotic genes. *See*: **TATA box**.

primary antibody In an **ELISA** or other immunological assay, the **antibody** that binds to the **target** molecule.

primary cell A cell or **cell line** taken directly from a living organism, which is not immortalized.

primary cell wall The cell wall layer formed during cell expansion. Plant cells possessing only primary walls may divide or undergo differentiation.

primary culture A culture started from cells, tissues or organs taken directly from organisms. A primary culture may be regarded as such until it is sub-cultured for the first time. It then becomes a cell line.

primary First in order of time or development.

primary germ layers *See*: **germ layer**.

primary growth 1. Apical **meristem**-derived growth; the tissues of a young plant. 2. **Explant** growth during the initial culture period.

primary immune response The immune response that occurs during the first encounter of a mammal with a given **antigen**.

primary meristem Meristem of the shoot or root tip giving rise to the primary plant body.

primary structure The linear sequence of **residues** making up a **polymer** such as a nucleic acid, **polysaccharide** or **protein**. *See*: **secondary structure, tertiary structure** and **quaternary structure**.

primary tissue A tissue that has differentiated from a primary **meristem**.

primary transcript The **RNA** molecule produced by **transcription** prior to any **post-transcriptional modifications**; also called a **pre-mRNA** in eukaryotes.

primer A short **oligonucleotide anneal**ed to a **template** of **single-stranded DNA**, providing a doubled stranded structure from which **DNA polymerase** will synthesize a new **DNA** strand to produce a duplex molecule.

primer walking A method for sequencing long (>1 kbp) cloned pieces of **DNA**. The initial sequencing reaction reveals the **sequence** of the first few hundred **nucleotides** of the cloned **DNA**. Using this, a new **primer** of about 20 nucleotides is synthesized, which is complementary to a sequence near the end of sequenced DNA, and used to sequence the next few hundred nucleotides of the cloned DNA. This procedure is repeated until the complete nucleotide sequence of the cloned DNA is determined.

primordium A group of cells which gives rise to an organ.

primosome A **protein**-replication complex that catalyses the initiation of synthesis of **Okazaki fragments** during discontinuous replication of **DNA**. It involves **DNA primase** and **DNA helicase** activities.

prion *See*: **proteinaceous infectious particle**.

probability The frequency of occurrence of an event.

proband The individual in a family in whom an inherited **trait** is first identified.

probe A labelled **DNA** or **RNA** sequence used to detect the presence of a **complementary** sequence by **hybridization** with a **nucleic acid** sample.

procambium A primary **meristem** that gives rise to primary vascular tissues and, in most woody plants, to the **vascular cambium**.

procaryote *See*: **prokaryote, prokaryotic**.

procaryotic (adj.) *See*: **procaryote**.

processed pseudo-gene A copy of a functional gene which has no promoter, no introns and which, consequently, is not itself transcribed.

production environment All input-output relationships, over time, at a particular location. The relationships include biological, climatic, economic, social, cultural and political factors, which combine to determine the productive potential of a particular enterprise. Production environments are classified as *high-*, *medium-* and *low-input*.

production traits Characteristics of animals, such as the quantity or quality of the milk, meat, fibre, eggs, work, etc., they (or their progeny) produce, which contribute directly to the value of the animals for the farmer, and that are identifiable or measurable at the individual level.

productivity The amount of economically significant product generated within a given period of time from a specified quantity of resource.

pro-embryo A group of cells arising from the division of the fertilized **egg** cell or **somatic embryo** before those cells which are to become the **embryo** are recognizable.

progeny Synonym of **offspring**.

progeny testing With respect to discrete loci, the inference of the **allelic** state of an individual from the pattern of **segregation** among its **offspring**. For a quantitative trait, the use of **progeny** performance to estimate the **breeding value** of an individual.

progesterone A **hormone** produced primarily by the corpus luteum, but also by the placenta. Its function in mammals is to prepare the inner lining of the uterus for the implantation of a fertilized egg. Also made by non-placental animals, including fish.

programmed cell death *See*: **apoptosis, p53 gene**.

prokaryote A member of the large group of organisms, including bacteria and blue-green algae, in which the **chromosome** is not enclosed within a **nucleus**, but instead exists as a linear or circular strand. Prokaryotes do not undergo **meiosis** and do not have functional **organelles** such as mitochondria and **chloroplasts**. *See*: **eukaryote**.

prolactin A **hormone**, produced by the anterior pituitary gland, that stimulates and controls lactation in mammals.

proliferation Increase by frequent and repeated reproduction; growth by **cell** division.

pro-meristem The embryonic **meristem** that is the source of **organ** initials or foundation cells.

promoter 1. A short **DNA** sequence, usually **upstream** of (5' to) the relevant **coding sequence**, to which **RNA polymerase** binds before initiating **transcription**. This binding aligns the **RNA** polymerase so that transcription will initiate at a specific site. The **nucleotide** sequence of the promoter determines the nature of the **enzyme** that attaches to it and the rate of **RNA** synthesis. 2. A chemical substance that enhances the **transformation** of benign cells into cancerous cells. *See*: **constitutive promoter**.

promoter sequence *See*: **promoter** (1).

pro-nuclear micro-injection An early, low success-rate method to achieve **transgenesis** in animals, involving the micro-injection of many **gene** copies into one of the two **pro-nuclei** of a fertilized egg. Now being replaced by micro-injection into a culture of cloned embryos produced by nuclear transfer, which can be tested for expression of the **transgene** before transfer to recipient females.

pro-nucleus Either one of the two **haploid gamete** nuclei, just prior to their fusion in the fertilized **egg**.

proofreading The scanning of newly-synthesized **DNA** for structural defects, such as mismatched base pairs. A functional activity of most **DNA polymerases.**

propagation The duplication of a whole plant from a range of vegetative materials; adapted for ***in vitro*** culture as **micropropagation.**

propagule Any structure capable of giving rise to a new plant by asexual or sexual reproduction, including bulbils, leafbuds, etc.

pro-phage The **genome** of a **bacteriophage** integrated into the **chromosome** of a **lysogenic** bacterial cell, and replicated along with its host chromosome.

prophase The first stage of nuclear division. The stage during which **chromosome pairing** occurs in the first division of **meiosis** (*see*: **leptonema, zygonema, pachynema, diplonema, diakinesis**). In **mitosis** and the second division of meiosis, the chromosomes shorten and thicken as a result of coiling.

protamine A class of small basic **proteins** that replace the **histones** in the chromosomes of some **sperm** cells.

protease An enzyme that catalyses the **hydroly-**

sis of **proteins**, cleaving the **peptide** bonds that link **amino acids** in **protein** molecules. *Synonym*: peptidase.

protein A macromolecule composed of one or more **polypeptides**, each comprising a chain of **amino acids** linked by **peptide** bonds.

protein crystallization The production of a pure preparation of a **protein**. In this form, the three-dimensional structure of the molecule can be determined.

protein drug *See*: **therapeutic agent**.

protein engineering Generating **proteins** with modified structures that confer novel properties such as higher catalytic **specificity** or thermal stability.

protein kinase An enzyme that catalyses the addition of a phosphate group(s) to a **protein** molecule at the sites of serine, threonine or tyrosine residues.

protein metabolic step One step in the chain of reactions that take place in an organism and dictate the composition of that organism.

protein sequencing The process of determining the **amino acid** sequence of a **protein**. Usually achieved following initially partial **hydrolysis** of the **protein** into smaller **peptides** by enzymatic digestion.

protein synthesis The creation of **proteins** from their constituent **amino acids**, in accordance with the encoding gene **DNA** sequence.

proteinaceous infectious particle Believed to be the agent responsible for the class of diseases called spongiform encephalopathies, including **scrapie** in sheep, bovine spongiform encephalopathy (BSE; mad cow disease) in cattle and CJD in humans. It is an abnormal form of a brain **protein**, and has no detectable **nucleic acid** content. *Synonym*: **prion**.

proteolysis Enzymatic degradation of a **protein**.

proteolytic Having the ability to degrade **protein** molecules.

proteome The complete complement of **proteins** made by a given **species** in all its tissues and growth stages.

proteomics An approach that seeks to identify and characterize complete sets of **protein**, and

protein-protein interactions in a given species. *See*: **proteome, genomics**.

protoclone Regenerated plant derived from **protoplast culture** or a single **colony** derived from protoplasts in culture.

protocol The step-by-step experiments proposed to describe or solve a scientific problem, or the defined steps of a specific procedure.

protocorm A tuberous structure, formed following the **germination** of orchid seeds, from which develops a complete plant. It develops from an unorganized embryo in the seed, comprising only a few hundred cells. In culture, vegetative **explants** of several orchid species form round, smooth protocorms which can be multiplied indefinitely or induced to regenerate into a whole plant.

protoderm A primary **meristem tissue** that gives rise to epidermis.

protogyny The condition in which the female reproductive organs (**carpels**) of a flower mature before the male ones (**stamens**), thereby ensuring that self-fertilization does not occur.

protomeristem *See*: **pro-meristem.**

proto-oncogene A normal gene that can be mutated to an **oncogene**. *Synonym*: **cellular oncogene.**

protoplasm The essential, complex living substance of cells, upon which all vital functions of nutrition, secretion, growth and **reproduction** depend.

protoplast A bacterial or plant cell for which the **cell wall** has been removed either chemically or enzymatically, leaving its **cytoplasm** enveloped by a peripheral membrane. Protoplasts are spherical and smaller than the elongate, angular shaped and often vacuolated cells from which they have been released.

protoplast culture The culturing *in vitro* of plant **protoplasts**. Where protoplasts can be regenerated into whole plants, they represent an attractive target for genetic manipulation.

protoplast fusion The induced or spontaneous coalescence of two or more **protoplasts** of the same or different species origin. Where fused protoplasts can be regenerated into whole plants, the opportunity exists for the creation of novel genomic combinations. *See*: **cybrid**.

prototroph A nutritionally independent cell. *Opposite*: **auxotroph**.

pro-toxin A latent, non-active precursor form of a **toxin**.

protozoan (pl.: protozoa) A microscopic, single-cell organism.

protruding end *See*: **extension**.

provenance The geographical and/or genetic origin of an individual.

provirus A double stranded **DNA** copy of the single **RNA** strand of a **retrovirus**, which has been integrated into a **host** genome.

pseudo-affinity chromatography A chromatographic technique in which a **ligand** is immobilized selectively to retain enzymes or other **proteins**.

pseudo-autosomal region A section at one end of the X and Y chromosomes for which there is sufficient **homology** that there is **synapsis** between them during **meiosis**.

pseudocarp A fruit that incorporates, in addition to the **ovary** wall, other parts of the flower, such as the **receptacle** (e.g. strawberry). *Synonym*: **false fruit**.

pseudogene An incomplete or mutated copy of a gene which is not transcribed because it lacks a continuous **open reading frame**. Those that lack **introns** are called processed pseudogenes and are most likely **cDNA** copies synthesized from **mRNA** by reverse transcriptase.

***Pseudomonas* spp.** A widely distributed Gram-negative bacterial genus. Many of the soil forms produce a **pigment** that fluoresces under ultraviolet light, hence the descriptive term fluorescent *Pseudomonas*.

P-site Abbreviation for **peptidyl-tRNA binding site**.

psychrophile A microorganism that can grow at temperatures below 30 °C and as low as 0 °C. *See*: **mesophile, thermophile**.

PUC A widely used **plasmid**, containing as a **marker** a galactosidase gene.

pulsed-field gel electrophoresis (Abbreviation: PFGE). A procedure used to separate very large (50 kbp to several Mbp) **DNA** molecules by alternating the direction of electric current in a pulsed manner across a gel.

punctuated equilibrium The occurrence of **speciation** events in bursts, separated by long intervals of **species** stability.

pure line A strain in which all members are genetically nearly identical and are indistinguishable by **phenotype**. Usually created by repeated generations of self-fertilization or close inbreeding.

purification tag *See*: **affinity tag.**

purine A double-ring, nitrogen-containing **base** present **in nucleic acids. Adenine** (A) and **guanine** (G) are the two purines normally present in **DNA** and **RNA** molecules.

PVP 1. Abbreviation for **polyvinylpyrrolidone**. 2. Abbreviation for **plant variety protection**.

PVR Abbreviation for **plant variety rights**.

PWP Abbreviation for **permanent wilting point**.

pyrethrins Active constituents of pyrethrum (*Tanacetum cinerariifolium*) flowers, used as insecticides.

pyrimidine A single-ring, nitrogen-containing **base** present in **nucleic acids. Cytosine** (C) and **thymine** (T) are present in **DNA**, whereas **uracil** (U) replaces T in **RNA**. Thymine is a synonym for 5-methyluracil.

pyrogen Bacterial substance that causes fever in mammals.

pyrophosphate A phosphate ion **dimer**; may be released on **hydrolysis** of **ATP**.

q Denotes the longer of the two **chromosome** arms, e.g. human 10q is the long arm of human chromosome 10.

q-beta replicase A viral **RNA polymerase** secreted by a **bacteriophage** that infects *E. coli*. It has the property of being able to copy **RNA** sequences at a rapid rate.

QSAR Abbreviation for **quantitative structure-activity relationship**.

QTL Abbreviation for **quantitative trait locus**.

quadrivalent A chromosome configuration visible in late **prophase** and **metaphase** of the first meiotic division, where four **chromosomes** are linked by chiasmata. Can occur in **autotetraploids** when four **homologous** chromosomes pair, or in **diploids** as a result of heterozygosity for a reciprocal translocation between two non-homologous chromosomes.

quadruplex The inheritance of alleles in **autotetraploids**. A **genotype** *AAAa* will produce gametes *AA*, *Aa* in the ratio 3:1.

qualitative trait A trait that shows **discontinu-**

ous variation - i.e. individuals can be assigned to one of a small number of discrete classes.

quantitative genetics The area of genetics concerned with the **inheritance** of **quantitative traits** that show continuous variation, as opposed to **qualitative traits**. Since many of the critical targets in both plant and animal **breeding** are of this type, most practical improvement programs involve the application of quantitative genetics.

quantitative inheritance Inheritance of measurable traits that depend on the cumulative action of many genes and/or involve a significant proportion of non-genetic determination.

quantitative structure-activity relationship (Abbreviation QSAR). A computer modelling technique that enables the prediction of the likely activity of a molecule before it is synthesized. QSAR analysis relies on recognizing associations of molecular structures and activity from historical data.

quantitative trait A measurable **trait** that shows **continuous variation** (e.g. height, weight, colour intensity, etc.) - i.e. the **population** cannot be classified into a few discrete classes.

quantitative trait locus (Abbreviation: QTL). A locus where **allelic variation** is associated with variation in a **quantitative trait**. The presence of a QTL is inferred from genetic mapping, where the total variation is partitioned into components linked to a number of discrete **chromosome** regions.

quantum speciation The rapid formation of new species, primarily by **genetic drift**.

quarantine Isolation for a period after arrival in a new location, to allow any pre-existing disease symptoms to appear. Used in the context of regulations restricting the sale or shipment of living organisms, usually to prevent disease or pest invasion of an area.

quaternary structure A level of **protein** structure where several individual molecules assemble together and form a functional cluster. A classic example is haemoglobin, a complex of four myoglobin-like units. *See*: **tertiary structure**.

quiescent A temporary suspension or reduction in the rate of activity or growth, while retaining the potential to resume prior activity. Applies particularly to **cell** division. *See:* **dormancy**.

R

R **genes** A class of plant genes conferring **resistance** to a specific **strain** (or group of strains) of a particular **pathogen**. Their primary function is to sense the presence of the pathogen and to trigger the defence pathways in the plant. *R* genes have been cloned from a number of plant species.

$\mathbf{R_1}$ The first-generation **offspring** of a **recombinant** (genetically modified) organism. Not standard terminology. *See*: $\mathbf{T_0}$, $\mathbf{T_1}$, **and** $\mathbf{T_2}$.

race A distinguishable group of organisms of a particular species. Criteria for distinctness can be one or a combination of geographic, ecological, physiological, morphological, genetic and karyotypic factors.

raceme An **inflorescence** in which the main axis is elongated but the flowers are borne on **pedicels** that are about equal in length.

rachilla Shortened axis of a **spikelet**.

rachis Main axis of a **spike**; axis of fern leaf (frond) from which pinnae arise; in compound leaves, the **extension** of the **petiole** corresponding to the midrib of an entire leaf.

radiation hybrid cell panel (Abbreviation: RH). A **somatic cell hybrid panel** in which the chromosomes from the species of interest have been fragmented by **irradiation** prior to **cell fusion**. The resultant small fragments of chromosomes greatly increase the power of **physical mapping** in the species of interest.

radicle The portion of the plant **embryo** which develops into the primary root.

radioimmunoassay (Abbreviation: RIA). An **assay** based on the use of a radioactively labelled antibody, where the amount of radiation detected indicates the amount of target substance present in the sample.

radioisotope An unstable **isotope** that emits ionizing radiation. *Synonym*: radioactive isotope.

raft culture *See*: **nurse culture**.

ramet An individual member of a clone, descended from the **ortet**.

random amplified polymorphic DNA (Abbreviation: RAPD). A **PCR**-based genotyping technique in which genomic **template** is amplified with single, short (usually 10-mer) randomly chosen **primers**. Typical patterns consist of a small number of amplified products of up to 2 kbp in length, which are separated by electrophoresis.

random genetic drift *See*: **genetic drift**.

random mutagenesis A non-directed change of one or more **nucleotide** pairs in a **DNA** molecule.

random primer method A method for **labelling DNA** probes, mainly for **Southern hybridization** experiments. A mixture of short oligonucleotides is hybridized to a **single-stranded DNA** probe. In the presence of **DNA polymerase** and **deoxyribonucleotides** - one of which is labelled - DNA synthesis then generates labelled copies of **probe** DNA.

RAPD Abbreviation for **random amplified polymorphic DNA**.

rate-limiting enzyme The enzyme whose activity controls the output of final product from a multi-enzyme metabolic pathway.

rational drug design A systematic method of creating compounds by analysing their structure, function and stereochemical interactions.

reading frame The reading frame defines which sets of three **nucleotides** are read as triplets, and hence as **codons**, in **DNA transcription**. The start point is usually determined by the **initiation codon**, AUG. Thus the sequence AUGGCAAAA would be read as AUG/GCA/AAA not as A/UGC/CAA/AA. *See*: **open reading frame**.

read-through Transcription or **translation** that proceeds beyond the normal stopping point because of the absence of the usual **transcription** or translation **termination signal** of a gene.

recA A **protein**, found in most bacteria, that is essential for **DNA repair** and **DNA** recombination.

recalcitrant Of seeds, unable to survive drying and subsequent storage at low temperature. *See*: **field gene bank**.

receptacle Enlarged end of the **pedicel** or **peduncle**, to which other flower parts are attached.

receptor A trans-membrane **protein** located in the **plasma** membrane that can bind with a **ligand** on the extracellular surface, as a result of which it induces a change in activity on the cytoplasmic surface. More generally, a site in a molecule that allows the **binding** of a ligand.

receptor-binding screening A biotechnology-based method for **drug** discovery, which relies on the fact that many drugs act by binding to specific **receptors** on or in cells. Since receptors *in vivo* bind to hormones or to other cells, and thereby control the cell's behaviour, a receptor bound with a drug will likely affect the normal activity of the cell.

recessive allele Allelic state of a gene, where homozygosity is required for the expression of the relevant **phenotype**. *Opposite*: **dominant allele**.

recessive Describing an allele whose effect with respect to a particular **trait** is not evident in **heterozygotes**. *Opposite*: **dominant**.

recessive oncogene A single copy of this gene

is sufficient to suppress **cell** proliferation; the loss of both copies of the gene contributes to cancer formation. *Synonym*: **anti-oncogene** recessive-acting oncogene. *See*: **oncogene**.

recessive-acting oncogene *See*: **recessive oncogene**.

reciprocating shaker A platform **shaker** used for agitating culture flasks, with a back and forth action at variable speeds.

recognition sequence Synonym of **recognition site**.

recognition site A nucleotide sequence, typically 4-8bp long and often palindromic, that is recognized by, and at which a **restriction endonuclease** binds to the **DNA**. For some restriction endonucleases, the presence of a methylated **residue** within the recognition site abolishes recognition. *Synonym*: **recognition sequence**; **restriction site**.

recombinant A term used in both classical and molecular genetics. 1. In classical genetics: An organism or cell that is the result of meiotic **recombination**. 2. In molecular genetics: A **hybrid** molecule made up of **DNA** obtained from different organisms. Typically used as an adjective, e.g. **recombinant DNA**.

recombinant DNA technology A set of techniques for manipulating **DNA**, including: the identification and **cloning** of genes; the study of the expression of cloned genes; and the production of large quantities of **gene** product.

recombinant DNA The result of combining **DNA** fragments from different sources.

recombinant human (Abbreviation rh). A prefix denoting molecules made through the use of recombinant **DNA** technology.

recombinant protein A **protein** encoded by a cloned gene. *Synonym*: **heterologous protein**.

recombinant RNA RNA molecules joined *in vitro* by T4 **RNA** ligase.

recombinant toxin A single multifunctional toxic **protein** encoded by a recombinant gene.

recombinant vaccine A **vaccine** produced from a cloned gene.

recombinase A class of enzymes that are able to

alter the arrangement of **DNA** sequences in a **site-specific** way.

recombination fraction The proportion of **recombinant** (with respect to two loci) gametes arising from meiosis. **Linkage maps** are based on estimates of recombination fraction between all pair-wise combinations of loci. *See*: **map distance**. *Synonyms*: **recombination frequency, crossing-over unit**.

recombination frequency *Synonym*: **recombination fraction.**

recombination The production of a **DNA** molecule with segments derived from more than one parent **DNA** molecule. In **eukaryotes,** this is achieved by the reciprocal exchange of **DNA** between non-sister chromatids within an **homologous** pair of chromosomes during **prophase** of the first meiotic division.

recombinational hot spot A chromosomal region where recombination appears to occur more frequently than expected.

reconstructed cell A viable transformed cell resulting from genetic engineering.

reduction division The first division of **meiosis** in which the **chromosome** number is reduced from the somatic to the gametic number.

refugium (pl.: refugia) An area set aside to provide protection/escape from ecological consequences occurring elsewhere.

regeneration The growth of new tissues or organs to replace those injured or lost. In plant **tissue culture**, regeneration refers to the **development** of organs or **plantlets** from an **explant**. *See*: **conversion; micropropagation; organogenesis.**

regulator Substance regulating growth and **development** of cells, organs, etc.

regulatory gene A gene with the primary function of controlling the rate of synthesis of the products of one or several other genes or pathways.

regulatory sequence A **DNA** sequence involved in regulating the expression of a gene, e.g. a **promoter** or **operator** region (in the **DNA** molecule).

rejuvenation 1. Reversion from adult to juvenile stage. 2. The process of regular reproduction of **seed** stocks or collections in **gene banks**, in order to ensure continued viability.

relaxed circle plasmid *See*: **plasmid**.

relaxed circle *See*: **nicked circle**.

relaxed plasmid A **plasmid** that replicates independently of the bacterial **chromosome** and is present in 10-500 copies per cell.

release factor 1. A soluble **protein** that recognizes **termination codons in mRNAs** and terminates **translation** in response to these codons. 2. A hormone, produced by the hypothalamus, which stimulates the release of a **hormone** from the anterior pituitary **gland** into the bloodstream.

remediation The cleanup or **containment** of a hazardous waste disposal site to the satisfaction of the applicable regulatory agency. This can sometimes be accomplished with naturally occurring or engineered **microorganisms** or plants. *See*: **bioremediation**.

renaturation Of **DNA**, the reforming of two complementary molecules into a double-stranded structure, following heat or chemical induction of dissociation (denaturation). Of **protein**, the resumption of three-dimensional conformation, allowing the molecule to function normally. Denaturation of many **proteins** is irreversible, but denatured **DNA** molecules will renature readily under appropriate chemical and physical conditions.

rennin An enzyme, secreted by cells lining the stomach in mammals, responsible for the clotting of milk. Used in the manufacture of certain dairy products.

repeat unit A sequence of **nucleotides** that occurs repeatedly, often in a head-to-tail arrangement (tandemly).

repetitive DNA DNA sequences that are present in a **genome** in many copies, some of it originating from retrotransposon activity. A substantial proportion of all eukaryotic genomes is composed of this class of **DNA**, whose biological function is uncertain. Sometimes referred to as 'junk **DNA**'.

replacement The addition of a cloned corrected copy of a defective gene. *See*: **homogenotization.**

replacement therapy The administration of metabolites, co-factors or hormones that are deficient as the result of a genetic disease.

replica plating Duplicating a **population** of bacterial colonies growing on **agar** medium in one Petri **plate** to agar medium in another Petri plate.

replicase A viral enzyme necessary for the **replication** of the **virus** in the **host** cell.

replication fork Y shaped structure associated with **DNA replication.** It represents the point at which the strands of **double-stranded DNA** are separated so that replication can proceed.

replication The *in vivo* synthesis of double-stranded **DNA** by copying from a single-stranded template.

replicative form (Abbreviation: RF). The molecular configuration of viral **nucleic acid** that is the **template** for replication in the host cell.

replicon The portion of a **DNA** molecule which can be replicated from a single **origin of replication.** Plasmids and the chromosomes of bacteria, phages and other viruses usually have a single origin of replication so that their entire genome constitutes a single replicon. Eukaryotic chromosomes have multiple origins of replication, so comprise several replicons. Also used to describe a **DNA** molecule capable of independent replication.

replisome The complete replication apparatus, present at a **replication fork,** that carries out the replication of **DNA.**

reporter gene A gene that encodes a product that can be readily assayed. Used as **a marker** to confirm the incorporation of a **transgene** into **a cell, organ** or **tissue,** and as a means of testing the efficiency of specific **promoters.**

repressible enzyme An enzyme whose activity can be diminished by the presence of a regulatory molecule.

repressible gene A gene whose expression can

be diminished or extinguished by the presence of a regulatory molecule.

repression Inhibition of **transcription** by preventing **RNA polymerase** from **binding** to the **transcription** initiation site.

repressor A **protein** which binds to a specific **DNA** sequence **upstream** from the **transcription** initiation site of a gene and prevents **RNA polymerase** from commencing **mRNA** synthesis.

reproduction 1. Sexual reproduction: the regular altermation of **meiosis** and **fertilization** which provides for the production of **offspring**. The main biological significance of **sexual** reproduction lies in the phenomenon of **recombination**. 2. **Asexual** or agamic reproduction: the development of a new individual from a single cell or group of cells in the absence of meiosis. *See*: **apomixis**.

repulsion A double **heterozygote** in which the **dominant** (or wild-type) allele at one locus and the **recessive** (or mutant) allele at a second linked locus occur on the same **chromosome** (genetic constitution *Ab/aB*). *Synonym*: *trans* configuration. *Opposite*: **coupling**, *cis* configuration.

residue 1. *See*: **polymer**. 2. Materials remaining after degradation and/or attempted removal, e.g. **pesticide** residues in food.

resistance factor A plasmid that confers **antibiotic** resistance to a bacterium.

resistance The ability to withstand **abiotic** (high temperature, drought etc.) or biotic (disease) **stress**, or a toxic substance. Often in the context of genetic **determination** of resistance.

rest period A physiological condition of **viable seeds**, buds or bulbs that prevents growth even in the presence of otherwise favourable environmental conditions. *Synonym*: **dormancy**.

restitution nucleus A single nucleus arising from a failure of nuclear division, either during **meiosis**, in which a **gamete** is formed with the unreduced **chromosome** number; or at **mitosis** to give a **cell** with a doubled chromosome number.

restriction endonuclease A class of enzymes that cut **DNA** after recognizing a specific **sequence**. The three types of restriction endonuclease are: I. Where the cut occurs within a random sequence at sites >1kbp from the **recognition sequence**, and has both **restriction** and **methylation** activities. II: Cuts within, or near a short, usually palindromic recognition sequence. A separate enzyme methylates the same recognition sequence. III: Cuts 24-26bp **downstream** from a short, asymmetrical recognition sequence, requires **ATP** and has both restriction and methylation activities. Type II **enzymes** are the class used for most molecular biology applications.

restriction enzyme Synonym of **restriction endonuclease.**

restriction exonuclease A class of enzymes that degrade **DNA** or **RNA**, starting from either the 5'- or the 3'-end.

restriction fragment A shortened **DNA** molecule generated by the cleavage of a larger molecule by one or more **restriction endonucleases.**

restriction fragment length polymorphism (Abbreviation: RFLP). A class of **genetic marker** based on the detection of variation in the length of **restriction fragments** generated when **DNA** is treated with **restriction endonucleases**. Differences in fragment lengths arise due to **genetic variation** with respect to the presence or absence of specific **recognition site**(s). RFLPs were initially detected by **Southern hybridization** but are now detected by **electrophoresis** of digested **PCR** product.

restriction map The linear arrangement of restriction **endonuclease recognition sites** along a **DNA** molecule.

restriction site Synonym of **recognition site**.

reticulocyte A slightly immature red blood cell.

retro-element Any of the integrated **retroviruses** or the **transposable elements** that resemble them.

retroposon A **transposable element** that moves via **reverse transcription** but lacks the **long terminal** repeat sequences necessary for autonomous transposition. Much of the **repetitive DNA** that makes up a large proportion of eukaryotic genomes consists of silenced (i.e. inactive) retroposons. *Synonym*: retrotransposon.

retroviral vectors Gene transfer systems based on viruses that have **RNA** as their genetic material.

retrovirus A class of eukaryotic **RNA** viruses that, by using **reverse transcription**, can form **double-stranded DNA** copies of their genomes, which can integrate into the chromosomes of an infected cell. Pathogenic retroviruses include HIV and the causative agents of many vertebrate animal cancers.

reversal transfer Transfer of a culture from a callus-supporting medium to a shoot-inducing medium.

reverse genetics *See*: **positional cloning.**

reverse mutation *See*: **reversion**.

reverse transcriptase An enzyme that uses an **RNA** molecule as a **template** for the synthesis of a **complementary DNA** strand. *Synonym*: **RNA**-dependent **DNA** polymerase.

reverse transcription The synthesis of **DNA** from a **template** of **RNA**, accomplished by **reverse transcriptase**.

reversion Restitution of a **mutant** gene to the wild-type condition, or at least to a form that gives the wild-type phenotype; more generally, the appearance of a **trait** expressed by a remote ancestor. *Synonym*: reverse mutation.

RF Abbreviation for **replicative form**.

RFLP Abbreviation for **restriction fragment length polymorphism.**

rh Abbreviation for **recombinant human**.

rhizobacterium A microorganism whose natural habitat is near, on, or in, plant roots.

Rhizobium (pl.: *Rhizobia*) Prokaryotic **species** which are able to establish a symbiotic relationship with leguminous plants, as a result of which elemental nitrogen is fixed or converted to ammonia. *See*: **nitrogen fixation.**

rhizosphere The soil region in the immediate vicinity of growing plant roots.

Ri plasmid A class of large conjugative **plas-**

mids found in the soil **bacterium** *Agrobacterium rhizogenes,* which can infect certain plants and cause **hairy root disease**. Like **Ti plasmids**, Ri plasmids include sequences that are transferred to plant cells and inserted into the plant's **DNA** as part of the **infection** process.

RIA Abbreviation for **radioimmunoassay.**

ribonuclease (Abbreviation: **RNAse**). Any enzyme that catalyses the **hydrolysis** of **RNA**.

ribonucleic acid (Abbreviation: **RNA**). An organic acid **polymer** composed of **adenosine, guanosine, cytidine** and **uridine ribonucleotides**. The genetic material of some viruses, but more generally is the molecule, derived from **DNA** by **transcription**, that either carries information (**messenger RNA**), provides sub-cellular structure (**ribosomal RNA**), transports **amino acids** (**transfer RNA**), or facilitates the biochemical modification of itself or other **RNA** molecules.

ribonucleoside *See*: **nucleoside.**

ribonucleotide *See*: **nucleotide.**

ribose A monosaccharide found in all ribonucleosides, ribonucleotides and **RNA**. Its close analogue, 2-deoxy**ribose,** is similarly found in all deoxyribonucleosides, deoxyribonucleotides and **DNA.**

ribosomal binding site A sequence of nucleotides near the 5' end of a bacterial **mRNA** molecule that facilitates the binding of the **mRNA** to the small ribosomal sub-unit. Also called the **Shine-Delgarno sequence**.

ribosomal DNA The coding locus for **ribosomal RNA**. This is generally a large and complex locus, typically composed of a large number of repeat units, separated from one another by the **intergenic spacer**. A **repeat unit** comprises a gene copy for each individual ribosomal **RNA** component, separated from one another by the **internal transcribed spacer**.

ribosomal RNA (Abbreviation: rRNA). The **RNA** molecules that are essential structural and functional components of **ribosomes,** where **protein** synthesis occurs. Different classes of **rRNA** molecule are identified by their sedimentation (S) values. *E. coli* ribosomes contain one 16S rRNA molecule (1541 **nucleotides** long) in one (small) ribosomal sub-unit, and a 23S rRNA (2904 nucleotides) and a 5S rRNA (120 nucleotides) in the other (large) sub-unit. These three rRNA molecules are synthesized as part of a large precursor molecule which also contains the sequences of a number of **tRNAs**. Special processing enzymes **cleave** this large precursor to generate the functional molecules. Constitutes about 80% of total cellular RNA.

ribosome The sub-cellular structure that contains both **RNA** and **protein** molecules and is the site for the **translation** of m**RNA** into **protein**. Ribosomes comprise large and small subunits.

ribosome-inactivating protein (Abbreviation: RIP). A class of plant **proteins** that inhibit normal **ribosome** function, and are thus highly toxic. Type 1 RIPs consist of single **polypeptide** chain **proteins**; type 2 (e.g. ricin) consist of two **proteins** linked by a **disulphide bridge**, one the **toxin** and the other a **lectin** that attaches to recognition sites on a **target** cell.

ribozyme An **RNA** molecule that can catalyse chemical cleavage of itself or of other **RNAs**. *Synonyms*: **catalytic RNA, gene shears**.

ribulose A keto-pentose sugar ($C_5H_{11}O_5$) involved in the carbon dioxide fixation pathway of **photosynthesis**.

ribulose biphosphate (Abbreviation: RuBP). A five-carbon sugar combined with carbon dioxide to form a six-carbon intermediate in the first stage of the dark reaction of photosynthesis.

rinderpest Cattle plague; a viral **infection** of cattle, sheep and goats.

RIP Abbreviation for **ribosome-inactivating protein**.

risk analysis A process consisting of three components: **risk assessment**, **risk management** and **risk communication** performed to understand the nature of unwanted, negative consequences to human and animal health, or the environment.

risk assessment a scientifically based process consisting of the following steps: i) hazard identification; ii) hazard characterization; iii) exposure assessment; and iv) risk characterization.

risk communication The interactive exchange of information and opinions throughout the risk analysis process concerning hazards and risks, risk-related factors and risk perceptions, among risk assessors, risk managers, consumers, industry, the academic community and other interested parties, including the explanation of risk assessment findings and the basis of risk management decisions.

risk management The process, distinct from **risk assessment**, of weighing policy alternatives, in consultation with all interested parties, considering risk assessment and other factors relevant for the health protection of consumers and for the promotion of fair trade practices, and, if needed, selecting appropriate prevention and control options.

R-loops Single-stranded **DNA** regions in **RNA-DNA** hybrids formed ***in vitro*** under conditions where **RNA-DNA** duplexes are more stable than **DNA-DNA** duplexes.

RNA Abbreviation for **ribonucleic acid.**

RNA editing Post-transcriptional processes that alter the information encoded in **RNAs**.

RNA polymerase A polymerase **enzyme** that catalyses the synthesis of **RNA** from a **DNA template.**

RNAase Abbreviation for **ribonuclease**.

RNA-dependent DNA polymerase *See*: **reverse transcriptase.**

RNase Abbreviation for **ribonuclease.**

***rol* genes** A family of genes, present on the **Ri plasmid** of *Agrobacterium rhizogenes*, that when transferred to a plant upon **infection** by the bacterium, induce the formation of roots. Used as a means of **root induction** on different species and cultivars of micropropagated fruit trees.

root apex The apical **meristem** of a root; very similar to the shoot apical meristem in that it forms the three meristematic areas: the **protoderm** (develops into the epidermis); the **procambium** (the stele); and the growth meristem (the cortex).

root cap A mass of reinforced cells covering and protecting the apical **meristem** of a root.

root culture The culture of isolated apical or lateral root tips to produce *in vitro* root systems with indeterminate growth habits. Used to study mycorrhizal, symbiotic and plant-parasitic relationships.

root cutting Cutting made from sections of roots alone.

root hairs Outgrowths from epidermal **cell walls** of the **root**, specialized for water and nutrient absorption.

root nodule A small round mass of cells attached to the roots of leguminous plants, containing symbiotic nitrogen-fixing bacteroids, particularly ***Rhizobium*** spp.

root The descending axis of a plant, normally below ground, which serves to anchor the plant and to absorb and conduct water and mineral nutrients.

root tuber Thickened **root** that stores carbohydrates.

root zone The volume of soil or growing medium containing the roots of a plant. In soil science, the depth of the soil profile in which roots are normally found.

rootstock The trunk or **root** material to which buds or **scions** are inserted in grafting. *See*: **stock**.

rotary shaker Rotating apparatus with a platform on which liquid **media** or cultures can be continuously shaken.

Roundup-readyä Describing **transgenic** crop varieties that carry the bacterial gene which detoxifies the **herbicide glyphosate**, thereby making them resistant to its application.

rRNA Abbreviation for **ribosomal RNA**.

RuBP Abbreviation for **ribulose biphosphate**.

ruminant Animal having a rumen - a large digestive sac in which fibrous plant material is fermented by commensal microbes, prior to its digestion in a "true" stomach (the *aboma-*

sum). Common farm ruminants are cattle and sheep.

runner A lateral **stem** that grows horizontally along the ground surface and gives rise to new plants either from axillary or terminal buds. *Synonym*: stolon.

rust A generic descriptor for various serious fungal plant pathogens, which infect the leaves and stems of crops. The appearance of spores is reminiscent of metallic rust, although the colour varies, according to species, from yellow to reddish-brown.

S

S phase The phase in the **cell cycle** during which **DNA** synthesis occurs.

S_1 mapping A method to characterise **post-transcriptional modifications** in **RNA** (removal of **introns** etc.) by hybridizing **RNA** with **single-stranded DNA** and treating with **S_1 nuclease**.

S_1 nuclease An enzyme obtained from the filamentous fungus *Aspergillus oryzae* which specifically degrades **RNA** or **single-stranded DNA** into its constituent mononucleotides, and cleaves nicked **double-stranded DNA** at the **nick**.

saccharifaction Following liquefaction, the **hydrolysis** of **polysaccharides** by glucoamylase to maltose and glucose.

saline resistance Synonym for **salt tolerance**.

Salmonella A genus of rod-shaped, Gram-negative bacteria that are a common cause of food poisoning.

salt tolerance The ability of a plant in soil or in culture to withstand a concentration of common salt (sodium chloride) which is damaging or lethal to most other plants. Breeding and selection for increased tolerance and **resistance** in crop plants is of great current interest. *Synonym*: saline resistance. An organism with extreme salt tolerance is a **halophyte**.

sap Fluid content of the **xylem** and **phloem cells** of plants. Fluid content of the **vacuole** generally referred to as cell sap.

saprophyte An organism (generally a fungus) that depends on dead plant or animal **tissue** for its source of nutrition and metabolic energy.

satellite DNA Highly **repetitive DNA** in plant and animal genomes, consisting of millions of copies of sequences typically in the range 5-500 bp long. Thousands of copies occur tandemly (head-to-tail) at each of many sites. It can be isolated from the rest of the genomic **DNA** by density gradient centrifugation.

satellite RNA A small, self-splicing **RNA** molecule that accompanies several plant viruses, including tobacco ringspot virus. *Synonym*: viroid.

SC Abbreviation for **synaptonemal complex**.

SCA Abbreviation for **specific combining ability**.

scaffold The central **protein**aceous core structure of condensed eukaryotic chromosomes. The scaffold is composed of non-**histone** chromosomal **proteins**.

scale up Conversion of a process, such as **fermentation** of a **microorganism**, from a small laboratory scale to a larger industrial scale.

scanning electron microscope (Abbreviation: SEM). An electron-beam-based microscope used to examine, in a three dimensional screen image, the surface structure of prepared specimens.

SCAR Abbreviation for **sequence characterized amplified region.**

scarification The chemical or physical treatment given to certain **seeds** having hard, impermeable seed coats in order to puncture or weaken the seed coat sufficiently to permit water uptake and germination.

SCE Abbreviation for **sister chromatid ex-**

change.

scion A twig or **bud** used for **grafting** onto another plant or **rootstock**.

scion-stock interaction The effect of a **rootstock** on a **scion** (and *vice versa*) in which a particular scion grafted onto a specific **s** performs differently than it would either on its own roots or on a different rootstock.

sclerenchyma A strengthening **tissue** in plants, composed of cells with heavily lignified **cell walls**.

SCP Abbreviation for **single-cell protein**.

scrapie A spongiform encephalopathy disease of sheep. *See*: **proteinaceous infectious particle**.

screen Preliminary characterization of a sample collection on the basis of a set of simple established criteria (biochemical, anatomical, physiological, etc.). Often applied to the process of **selection** for specific purposes, such as for **disease resistance** or for improved agronomic performance in crop plants.

SDS Abbreviation for **sodium dodecyl sulphate**.

SDS-PAGE Abbreviation for **sodium dodecyl sulphate polyacrylamide gel electrophoresis**.

secondary antibody In an **ELISA** or other immunological **assay** system, the antibody designed to bind to the **primary antibody**, and to which a **label** is generally attached.

secondary cell wall The innermost layer of cell wall, giving rigidity to the cells. Characterized by its highly organized microfibrillar structure, and only formed in certain cells after cell elongation has ceased.

secondary growth Type of growth characterized by an increase in the thickness of **stem** and root, and resulting from the formation of secondary **vascular tissues** by the vascular **cambium**.

secondary immune response The rapid immune response that occurs during the second (and subsequent) encounters of the immune system of a mammal with a specific **antigen**. *See*: **primary immune response**.

secondary messenger A chemical compound

within a **cell** that is responsible for initiating the response to a signal from a chemical messenger (such as a hormone) that cannot enter the **target** cell itself.

secondary metabolism The production by living organisms of substances not essential for **primary** metabolic functions or physiology. Their role is associated with interaction with the environment, for example for defence, as **elicitors** or as attractants. Some of these have useful pharmacological or nutritional properties, while others are toxic.

secondary metabolite Product of **secondary metabolism**.

secondary oocyte *See*: **oocyte**.

secondary phloem Phloem tissue formed by the **vascular cambium** during secondary growth in a vascular plant.

secondary plant product *See*: **secondary metabolite**.

secondary root A branch or lateral root.

secondary spermatocyte *See*: **spermatocyte**.

secondary structure Localized three dimensional **conformations** adopted by **macromolecules**, in particular nucleic acids and polypeptides. These arise as a result of the action of non-covalent forces generated by interactions between **residues** which are brought into close contact with one another. Examples are alpha-helix regions and beta-pleated sheets in **proteins**, and hairpin loops in nucleic acids. *See*: **primary structure, tertiary structure, quaternary structure.**

secondary thickening Deposition of secondary **cell wall** materials which result in an increase in thickness in stems and roots.

secondary vascular tissue Vascular tissue (**xylem** and **phloem**) formed by the vascular cambium during secondary growth in a vascular plant.

secondary xylem *See*: **secondary vascular tissue**.

secretion The transport of a **molecule** from the inside of a **cell** through the cell membrane.

seed Botanically, the matured **ovule** without accessory parts. Colloquially, anything which may be sown; i.e. seed potatoes (which are vegetative tubers); seed of wheat (karyopses) etc.

seed storage proteins Proteins accumulated in large amounts in **protein** bodies within **seeds**. They act as a source of **amino acids** during germination. Of interest in biotechnology: 1. As a major source of human and animal nutritional protein. 2. As a model expression system. Since they are produced in large amounts relative to other proteins, and are stored in stable, compact bodies in the plant seed, it may be possible to engineer **transgenes** which are expressed in the same way as seed storage proteins - i.e. in large amounts and in a convenient form.

segment-polarity gene A gene that functions to define the anterior and posterior components of body segments in *Drosophila*.

segregant An individual derived from a **cross** between two unlike parents.

segregation For genes, the separation of allele pairs from one another and their resulting assortment into different cells at **meiosis**. For chromosomes, the separation and re-assortment of the two **homologues** in **anaphase** of the first meiotic division. For individuals, the occurrence of different genotypes and/or phenotypes among offspring, resulting from **chromosome** or **allele** separation in their **heterozygous** parents.

selectable Having a gene product that, when present, enables the identification and preferential **propagation** of a particular **genotype**. *See*: **reporter gene**.

selectable marker A gene whose expression allows the identification of a specific **trait** or **gene** in an organism.

selection 1. Differential survival and reproduction of **phenotypes**. 2. A system for either isolating or identifying specific **genotypes** in a mixed population.

selection coefficient A measure of the intensity of selection at a locus, commonly abbreviated as *s*. It represents the proportionate reduction in the gametic contribution of a particular **genotype**, compared with the (generally most favoured) standard genotype.

selection culture A selection based on difference(s) in environmental conditions or in **culture medium** composition, such that preferred **variant** cells or **cell lines** (presumptive or putative **mutants**) are favoured over other variants or the **wild type**.

selection differential The difference between the **mean** of the individuals selected to be parents and the mean of the overall **population**; it represents the average superiority of the selected parents; commonly abbreviated as *S*.

selection pressure The intensity of selection acting on a population of organisms or on cells in culture. Its effectiveness is measured in terms of differential survival and reproduction, and consequently in changes in **allele frequency** in a population.

selection response The difference between the mean of the individuals selected to be parents and the mean of their offspring. Predicted response is calculated as the product of **narrow-sense heritability** and **selection differential**.

self-incompatibility In plants, the inability of the **pollen** to fertilize ovules (female gametes) of the same plant.

self-replicating elements Extrachromosomal **DNA** elements that have origins of **replication** for the **initiation** of their own **DNA** synthesis.

self-sterility Synonym of **self-incompatibility**.

SEM Abbreviation for **scanning electron microscope**.

semen sexing Synonym of **sperm sexing**.

semi-conservative replication During **DNA** duplication, each strand of a parent **DNA** molecule acts as a **template** for the synthesis of a new **complementary** strand. Thus, one half of a pre-existing **DNA** molecule is conserved during each round of replication.

semi-continuous culture Cells in an actively dividing state which are maintained in culture by periodically draining off the medium and replenishing it with fresh **medium**.

semi-permeable membrane A natural or synthetic material which selectively allows the **passage** of certain ions or molecules.

semi-sterility The condition of partial fertility. Often associated with chromosomal aberrations or the result of **mutagenesis**.

senescence A late stage in the **development** of multicellular organisms, during which irreversible loss of function and degradation of biological components occur. The physiological ageing process in which cells and tissues deteriorate and finally die.

sense RNA The **RNA** transcript of the **coding strand DNA** (often represented as the (+)-strand). *Opposite*: **antisense RNA**. When both sense and antisense **transcripts** of a gene are present simultaneously, **gene silencing** is often the result.

sensitivity In diagnostic tests, the smallest amount of the **target** molecule that the **assay** can detect.

sepsis Destruction of **tissue** by pathogenic microorganisms or their toxins, especially through **infection** of a wound.

septate (adj.) *See*: **septum**.

septum A dividing wall or partition, which splits a structure into separate cells or compartments.

sequence characterized amplified region (Abbreviation: SCAR). A molecular **marker** obtained by the **conversion** to a **sequence-tagged site** of a single **random amplified polymorphic DNA** product.

sequence divergence The percent difference in the **nucleotide** sequence between related **nucleic acid** sequences, or in the **amino acid** sequence in a comparison between related **proteins**.

sequence hypothesis The concept that **genetic information** exists as a linear **DNA** code, and that **DNA** and gene product sequence are collinear.

sequence tandem repeat (Abbreviation: STR). *See*: **tandem repeat**.

sequence The linear order of **nucleotides** along a **DNA** or **RNA** molecule, and the process of obtaining this. **Genome** sequencing aims to generate the linear order of all nucleotides present in the nuclear DNA of an organism.

sequence-tagged site (Abbreviation: STS). Short unique **DNA** sequence (200-500 bp long) that can be amplified by **PCR** and is thus tagged to the site on the **chromosome** from which it was amplified.

serial division Splitting of excised shoot-tip material growing *in vitro,* in order to induce the development of greater numbers of **plantlets**.

serial float culture A technique whereby immature anthers are floated on a liquid medium, and continue their **development** through to the release of **pollen**.

serology The study of **serum** reactions between an **antigen** and its **antibody**. Mainly used to identify and distinguish between antigens, such as those specific to particular microorganisms or viruses.

serum albumin A globular **protein** obtained from blood and body fluids. Bovine and human serum albumins are abbreviated **BSA** and **HSA** respectively.

serum Blood **plasma** that has had its clotting factor removed.

sewage treatment A widespread biotechnological processes in developed economies. Methods vary widely, but all are designed for the biological break-down of human and animal waste in order to allow safe discharge into the environment.

sex chromosome Differentiated chromosome which is responsible for the **determination** of sex of the individual. For all mammals, a small number of flowering plants and many insects, female individuals carry a pair of X chromosomes, and males carry one X and one Y. For birds, reptiles and most amphibians, male individuals carry a pair of W chromosomes, and females carry one W and one Z. In some insects there is only one sex chromosome, X, and sex is determined by the number of these present. *Synonym*: allosome. *Opposite*: autosome.

sex determination Any method by which the distinction between males and females is established in a species, particularly at an early stage of foetal development.

sex duction The incorporation of bacterial genes into **F factors** and their subsequent transfer, by **conjugation**, to a recipient cell.

sex factor A bacterial **episome** (e.g. the F **plasmid** in *E. coli*) that enables the cell to be a donor of genetic material. The sex factor may be propagated in the **cytoplasm**, or it may be integrated into the bacterial **chromosome**.

sex hormones Steroid hormones that control sexual **development** in animals.

sex linkage Referring to genes present on one of the sex chromosomes, thus genetically linked to the sex of the individual.

sex mosaic Synonym of **gynandromorph**.

sexed embryos Embryos separated according to sex.

sex-influenced dominance The tendency for gene action to vary between the sexes within a species. For example, the presence of horns in some breeds of sheep appears to be **dominant** in males but **recessive** in females.

sex-limited Expression of a **trait** in only one sex; e.g. milk production in mammals; **egg** production in chickens.

sexual reproduction The process whereby two **gametes** fuse to form one fertilized **cell** (**zygote**).

shake culture An agitated suspension in culture providing adequate aeration for cells in the liquid medium. Usually achieved using platform **shakers**, or by constant stirring with a magnetic stirrer.

shaker A platform, with set or variable speed control, used to agitate vessels containing liquid cell cultures. Also described as a platform shaker.

shear Literally the sliding of one layer across another, with deformation and fracturing in the direction parallel to the movement. In the present context, used to describe 1. the forces that cells are subjected to in a bioreactor or a mechanical device used for cell breakage. 2.

the intentional or unintentional fragmentation of large **DNA** molecules, achieved commonly by passing a concentrated **DNA** solution through a hypodermic needle. This treatment generates random breaks in the **DNA**, and the average size of fragments can be manipulated by varying the bore of the needle.

Shine-Dalgarno sequence A conserved sequence of prokaryotic **mRNAs** that is **complementary** to a sequence near the 5' terminus of the 16S ribosomal **RNA** and is involved in the **initiation** of **translation**. *See*: **ribosomal binding site**.

shoot apex *See*: **shoot tip**.

shoot differentiation The development of growing points, leaf primordia and finally shoots from a shoot tip, axial bud, or callus surface.

shoot tip The terminal **bud** (0.1 - 1.0 mm) of a plant, which consists of the **apical meristem** (0.05 - 0.1 mm) and the immediately surrounding leaf primordia and developing leaves, and adjacent stem tissue. *Synonym*: **shoot apex**.

shoot-tip graft A **shoot tip** or **meristem tip** grafted onto a prepared seedling or micropropagated **rootstock** in culture. Meristem tip **grafting** is mainly used for *in vitro* **virus** elimination from *Citrus* spp. and other plants. *Synonym*: **micrograft**.

short interspersed nuclear element (Abbreviation: **SINE**). Families of short (150-300 bp), moderately **repetitive DNA** elements of eukaryotic genomes. They appear to be **DNA** copies of certain **tRNA** molecules, created presumably by the unintended action of **reverse transcriptase** during retroviral infection.

short-day plant A plant which will not flower until triggered to do so by exposure to one or a number of dark periods equal to or longer than its critical period. Other plant species are long-day and some are daylength neutral. **Genetic variation** for daylength **sensitivity** is present in many crop species.

shotgun genome sequencing A strategy for sequencing a whole **genome**, in which the genomic **DNA** is initially fragmented into pieces small enough to be sequenced. Specialized computer software is then used to piece together the individual sequences to create long contiguous tracts of sequenced **DNA**.

shuttle vector A **plasmid** capable of replicating in two different **host** organisms because it carries two different origins of **replication** and can therefore be used to transfer genes from one to the other. *Synonym*: **bifunctional vector**.

sib-mating The deliberate crossing of siblings. Generally done where **self-incompatibility** prevents the production of self-fertilized progeny.

siderophore A low molecular weight entity that binds very tightly to iron. Siderophores are synthesized by a variety of soil microorganisms to ensure that the **organism** is able to obtain sufficient amounts of iron from the environment.

sieve cell A long and slender **sieve element** in **vascular** plants, characterized by relatively unspecialized sieve areas and tapering end walls that lack **sieve plates**.

sieve element The **phloem** cell concerned with longitudinal conduction of food materials.

sieve plate Perforated wall area in a **sieve tube** element, through which strands connecting sieve tube protoplasts can pass.

sieve tube A tube within the **phloem** tissue of a plant, composed of joined **sieve elements**.

sigma factor The sub-unit of prokaryotic **RNA polymerases** responsible for the **initiation** of **transcription** at specific initiation **sequences**.

signal peptide *See*: **signal sequence**.

signal sequence A stretch of 15-30 amino acid residues at the N terminus of a **protein**, which is thought to enable the **protein** to be secreted (pass through a cell membrane). The signal sequence is removed as the **protein** is secreted. *Synonyms*: **signal peptide, leader peptide**.

signal transduction The biochemical events that conduct the signal of a **hormone** or **growth factor** from the cell exterior, through the cell membrane, and into the cytoplasm. This involves a number of molecules, including receptors, ligands and messengers.

signal-to-noise ratio A specifically produced response (signal) compared to the response level (noise) when no specific stimulus (activity) is present.

silencing Loss of **gene expression** either through an alteration in the **DNA sequence** of a structural gene, or its regulatory region; or because of interactions between its **transcript** and other **mRNAs** present in the cell (*see*: **antisense RNA**).

silent mutation *See*: **mutation**.

simple sequence repeat (Abbreviation: SSR). *See*: **microsatellite**.

SINE Abbreviation for **short interspersed nuclear element**.

single copy A gene or **DNA** sequence which occurs only once per (**haploid**) genome. Many structural genes are single copy.

single domain antibody *See*: **dAb**.

single node culture Culture of separate lateral **buds**, each carrying a piece of **stem** tissue.

single nucleotide polymorphism (Abbreviation: SNP). A genetic **marker** resulting from **variation** in **sequence** at a particular position within a **DNA** sequence. SNPs are commonly the result of **transition** changes (A for G, T for C), but also transversions (G or A for T or C) and single base deletions. Such variation is extensive throughout all genomes, and offers the particular advantage of being detectable without the need for **gel electrophoresis**.

single primer amplification reaction (Abbreviation: SPAR). A PCR-based genotyping technique in which genomic template is amplified with a single **primer**.

single-cell line *See*: **cell strain**.

single-cell protein (Abbreviation: SCP). **Protein** produced by microorganisms, particularly yeast. Used as either a feed or a food additive.

single-strand conformational polymorphism (Abbreviation: SSCP). A technique for detection of mutations in a defined **DNA** sequence.

Single-stranded polynucleotides are electrophoretically separated on non-denaturing gels. Intrachain base pairing results in a limited number of conformers stabilized by intrachain loops, and mutated **DNA** shows on **electrophoresis** an altered **assortment** of such conformers.

single-strand DNA binding protein A **protein** that coats **single-stranded DNA**, preventing **renaturation** and so maintaining the **DNA** in an extended state.

single-stranded DNA (Abbreviation: ssDNA). **DNA** molecules separated from their **complementary** strand, either by its absence or following denaturation.

single-stranded nucleic acid Nucleic acid molecules consisting of only one **polynucleotide** chain. The genomes of many viruses are **single-stranded DNA** molecules, as are most biologically effective RNAs. Many **RNA** molecules do include double-stranded regions formed by the intra-strand base-pairing of self-complementary sequences, and these determine the 3-dimensional shape (**conformation**) that they adopt *in vivo*.

sire Male animal chosen for breeding.

sister chromatid exchange (Abbreviation: SCE). Reciprocal interchanges of the two **chromatid** arms within a single **chromosome**.

site-specific A term used to describe any process or **enzyme** which acts at a defined sequence within a **DNA** or **RNA** molecule.

site-specific mutagenesis The induction of **mutations**, by molecular biology techniques, in one or more specific **nucleotides** within a defined **coding sequence** in order to create altered forms of the **gene** product. Used to define the **active sites** of **proteins** and for **protein engineering**.

sitosterol *See*: **phytosterol**.

six-base cutter Type II **restriction endonucleases** whose **recognition site** and cleavage site consist of a characteristic sequence of six **nucleotide** pairs. *See*: **four-base cutter**.

small nuclear ribonucleoprotein (Abbreviation: snRNP). A complex of **small nuclear RNA**

and nuclear **protein**, heavily involved in the post-transcriptional processing of m**RNA**, especially the removal of **introns**. snRNPs are a major component of **spliceosomes**.

small nuclear RNA (Abbreviation: sn**RNA**). **RNA** transcripts of 100-300 bp that associate with **proteins** to form **small nuclear ribonucleoprotein** particles. Most sn**RNAs** are components of the **spliceosomes**.

SNP Abbreviation for **single nucleotide polymorphism**.

snRNA Abbreviation for **small nuclear RNA**.

snRNP Abbreviation for **small nuclear ribonucleoprotein**.

sodium dodecyl sulphate (Abbreviation: SDS). A detergent used to solubilize **protein** and **DNA** from biological materials. Specific use in **sodium dodecyl sulphate polyacrylamide gel electrophoresis**.

sodium dodecyl sulphate polyacrylamide gel electrophoresis (Abbreviation: SDS-PAGE). A widely employed electrophoretic method for the separation of **proteins** from biological samples. The **sodium dodecyl sulphate** gives a uniform charge density to the surface of **proteins** or nucleic acids, so that their rate of migration through the gel is determined largely by their molecular weight.

soil amelioration The improvement of poor soils. Includes the fungal and bacterial break down of plant organic matter, to form humus; the release of minerals - such as phosphates - to the soil, making them available to plants; the fixation of nitrogen. Can sometimes include an element of **bioremediation**.

soil-less culture Growing plants in nutrient solution without soil. *Synonym*: **hydroponics**.

solid medium Nutrient medium solidified by the addition of a gelling agent, commonly agar.

somaclonal variation Epigenetic or genetic changes induced during the callus phase of plant cells cultured *in vitro*. Sometimes visible as changed **phenotype** in plants regenerated from culture.

somatic cell Cells not involved in sexual reproduction, i.e. not **germ** cells.

somatic cell embryogenesis The process of **differentiation** of **somatic embryos** either from **explant** cells (direct embryogenesis), or from callus generated from explants (indirect embryogenesis). *Synonym*: **asexual embryogenesis.**

somatic cell gene therapy The delivery of a **transgene**(s) to a **somatic tissue** in order to correct a physiological defect.

somatic cell hybrid panel A panel of cells created by **cell fusion**, typically involving a reference species (e.g. hamster) and the species of interest (e.g. sheep) with each member of the panel containing a different mixture of chromosomes from the two species. By relating the presence or absence of cloned fragments (via ***in situ* hybridization**) or **PCR** products to the presence or absence of particular chromosomes from the species of interest, such panels can be used for **physical mapping**.

somatic cell variant A somatic cell with unique characters not present in the other cells, and which could be selected for by an appropriate screen.

somatic embryo An organized **embryo**-like structure. Although morphologically similar to a zygotic embryo it is initiated from somatic plant cells. Under *in vitro* conditions, somatic embryos go through developmental processes similar to embryos of zygotic origin. Each somatic embryo is potentially capable of developing into a normal **plantlet**.

somatic hybridization Naturally occurring or induced fusion of somatic **protoplasts** or cells of two genetically different parents. The difference may be as wide as interspecific. Wide synthetic hybrids formed in this way (i.e. not via gametic fusion) are known as **cybrids**. Not all cybrids contain the full **genetic information** (nuclear and non-nuclear) of both parents.

somatic hypermutation The high frequency of **mutation** that occurs in the **gene** segments encoding the variable regions of **immuno-**

globulins during the **differentiation** of B **lymphocytes** into **antibody** producing **plasma** cells.

somatic reduction Halving of the chromosomal number of **somatic** cells; a possible method of producing "haploids" from somatic cells and calli by artificial means.

somatic Referring to cell types, structures and processes other than those associated with the **germ line**.

somatocrinin Growth hormone-releasing hormone. *See*: **growth hormone.**

somatostatin Growth hormone-inhibiting hormone. *See*: **growth hormone.**

somatotropin *See*: **growth hormone**.

sonication Disruption of cells or **DNA** molecules by high frequency sound waves.

SOS response The synthesis of a whole set of **DNA repair, recombination** and **replication proteins** in bacteria suffering severe **DNA damage** (e.g. following exposure to UV light).

source DNA The **DNA** from an organism that contains a **target** gene, and used as the starting material in a **cloning** experiment.

source organism A bacterium, plant or animal from which **DNA** is purified and used in a **cloning** experiment.

Southern blot A **nitrocellulose** or nylon membrane to which **DNA** fragments previously separated by gel **electrophoresis**, have been transferred by capillary action. *See:* **blot**.

Southern hybridization A procedure in which a cloned, labelled segment of **DNA** is hybridized to **DNA** restriction fragments on a **Southern blot.**

spacer sequence A **DNA** sequence separating neighbouring genes; spacer sequences are not usually transcribed.

SPAR Abbreviation for **single primer amplification reaction.**

sparger A device that introduces, into a bioreactor, air in the form of fine bubbles.

spatial autocorrelation statistics A set of statistical parameters aimed to depict the spatial (geographical) pattern of **genetic diversity** in a population.

speciation The evolutionary differentiation of a pre-existing species into one or more distinct species.

species A class of individuals capable of interbreeding, but which is reproductively isolated from other such groups having many characteristics in common. A somewhat arbitrary and sometimes blurred classification; but still quite useful in many situations.

specific combining ability (Abbreviation: SCA). A component of genetic **variance** calculable where a number of genotypes are intercrossed in all possible combinations. The SCA measures the **deviation** of the performance of a particular **cross** from the average general combining ability of its two parents.

specificity For diagnostic tests, the ability of a **probe** to react precisely and uniquely with its **target** molecule.

spent medium After sub-culture, medium which is discarded because it has been depleted of nutrients, dehydrated, or accumulated toxic metabolic products.

sperm Abbreviation for **spermatozoon**.

sperm competition Competition between different **spermatozoa** to fertilize the **egg cell** of a single female.

sperm sexing The separation of mammalian **sperm** into those bearing an **X chromosome** and those bearing a Y chromosome, in order to be able to produce, via **artificial insemination** or ***in vitro*** **fertilization**, animals of a specified sex. Methods for achieving this include the inactivation of X-bearing or Y-bearing sperm by antibodies recognizing sex-specific sperm surface peptides, and **fluorescence-activated cell sorting**.

spermatid Immature **spermatozoon**. One of the four cells formed at the end of the second meiotic division in **spermatogenesis**.

spermatocyte The premeiotic parental cell of the **spermatids**; the ***primary*** spermatocyte before the initiation of the first meiotic division; the ***secondary*** spermatocyte after completion of the first meiotic division, but before the initiation of the second division. *Synonym*: **sperm**

mother cell.

spermatogenesis The series of cell divisions in the testis as a result of which the formation and the **maturation** of the male **gametes** (i.e. sperm) are achieved.

spermatogonium (pl.: spermatogonia) Primordial male **germ cell**. These can either divide by **mitosis** to produce daughter cells, or enter a **growth phase** and differentiate into a primary **spermatocyte.**

spermatozoon (Abbreviation: sperm). (pl.: spermatozoa) The mature, mobile gametic cell of male animals, produced in the testis.

spharoblast Nodule of wood which can give rise to **adventitious** shoots with **juvenile** characteristics.

spheroplast (Alternative spelling: sphaeroplast). A microbial or plant cell from which most of the cell wall has been removed, usually by enzymatic treatment. Strictly, in a spheroplast, some of the cell wall remains, while in a **protoplast** the cell wall has been completely removed. In practice, the two words are often used interchangeably.

spike 1. An **inflorescence** in which the main axis is elongated and the flowers are sessile. 2. The deliberate addition of a known quantity of a known substance to an analytical sample, used to validate the analytical technique.

spikelet The unit of **inflorescence** in grasses, made up of a small group of florets.

spindle An intracellular fibrous structure, involved in the control of **chromosome** movement in **mitosis** and **meiosis.**

spliceosome A complex of **small nuclear ribonucleoproteins** and other **proteins** that assemble on an immature **mRNA** and catalyse the excision of an **intron**. *See*: **splicing.**

splicing 1. During the maturation of eukaryotic **mRNA,** the process that removes **intron** sequences and covalently joins **exon** sequences. *Synonym*: editing. 2. In **recombinant DNA** technology, the term refers to the **ligation** of two fragments of **DNA** together.

splicing junction The **DNA** sequence immediately surrounding the boundary between an

exon and an **intron**. There is a degree of sequence conservation in these regions, allowing the identification of introns in newly sequenced genes.

split gene In eukaryotes, the encoding **DNA** of many structural genes is made up of **exons** and **introns**. This commonly found pattern of interruption in the **coding sequence** is referred to as a 'split gene'.

spontaneous mutation A mutation occurring in the absence of any known **mutagen**.

sporangium (pl.: sporangia) A reproductive structure in plants that produces spores. A megasporangium produces megaspores, which give rise to the female gametophyte; in seed plants, it is represented by the **ovule**. A microsporangium produces microspores, which give rise to the male gametophyte; it is represented in **seed** plants by the **pollen** sac.

spore 1. A reproductive **cell** that develops into an individual without union with other cells; some spores such as meiospores are the product of the **germ** line, but others are **asexual** in nature. 2. A small, protected resting body, often synthesized by microorganisms when nutrient levels are low.

spore mother cell Synonym of **sporocyte**.

sporocyte A **diploid** germ line cell that is the parent of the four **haploid** spores generated by meiosis.

sporophyll A leaf that bears **sporangia**.

sporophyte The **diploid** generation in the life cycle of a plant, and that produces haploid **spores** by meiosis.

sport An individual plant, or portion thereof, showing a recognizably different **phenotype** from the parent, presumably as a result of spontaneous mutation. Novel traits displayed by some sports can become of great agricultural worth, but generally they are disadvantageous.

ssDNA Abbreviation for **single-stranded DNA**.

SSR Abbreviation for **simple sequence repeat**. *See*: **microsatellite**.

stacked genes Refers to the insertion of two or more genes into the genome of an organism.

An example would be a plant carrying a ***Bt*** **transgene** giving insect **resistance**, and a *bar* transgene giving resistance to a specific herbicide.

stages of culture (I-IV) *See*: **micropropagation**.

staggered cuts Symmetrically cleaved phosphodiester bonds that lie on both strands of double-stranded **DNA**, but are not opposite one another.

stamen Floral structure made up of an **anther** and a filament. The stamen is the male **organ** of a flower.

standard deviation A statistical measure of variability in a **population** of individuals or in a set of data.

standard error A statistical measure that indicates the predictive accuracy over all individuals of a **mean** value derived from a sample population.

starch The major plant **carbohydrate** storage substance, particularly but not exclusively found in seeds, and used both as food and feed source and for various industrial processes. A large water-insoluble heterogenous group of **polysaccharides**, consisting of various proportions of the two glucose **polymers**, **amylose** and **amylopectin**. Starch is broken down into simple metabolisable sugars *in vivo* by the action of **amylases**.

start codon The codon which specifies the first **amino acid** of a **polypeptide** chain and at which the **ribosome** starts the process of **translation**. In bacteria, this is either AUG (translated as *n*-formyl methionine) or, rarely, GUG (valine). In eukaryotes, it is always AUG and is translated as methionine. The start codon sets the **reading frame** for translation. *Synonym*: **initiation codon**.

starter culture **Microorganisms** that are deliberately added to foods to alter flavour, colour, texture, smell, or taste.

stationary culture A culture maintained without agitation.

stationary phase The plateau of the growth curve, during which **cell number** remains relatively constant, following the **logarithmic**

phase. *See*: **growth phases**.

steady state In a **continuous fermentation** process, the condition under which the number of cells removed with the **outflow** is exactly balanced by the number of newly synthesized cells.

stele The central **vascular** cylinder, inside the cortex, of roots and stems of higher plants.

stem cell An undifferentiated **somatic cell** that is capable of either division to give rise to daughter stem cells, or differentiating into any specialized cell type given the appropriate signals. Cultured stem cells are critical to the concept of **therapeutic cloning**.

stem The main body of the above-ground portion of a tree, shrub, herb or other plant; the ascending axis, whether above or below ground, of a plant.

sterile 1. Medium or object free of **viable** microorganisms (*see*: disinfect). 2. Incapable of producing viable gametes.

sterile room Dedicated space for the carrying out of activities that require **sterile** conditions. Can usually be achieved more economically with a **laminar air-flow cabinet**.

sterility Complete or partial failure of an individual to produce functional gametes or **viable** zygotes under a given set of environmental conditions.

sterilize 1. The elimination of microorganisms, using heat, irradiation, **filtration** or chemicals. 2. The operation of making an animal incapable of producing offspring.

Steward bottle Flask developed for the growth of cells and tissues in a liquid medium, in which they can be periodically submerged during rotation.

sticky end *See*: **extension**.

stigma Receptive portion of the **style**, to which **pollen** adheres.

stirred-tank fermenter A growth vessel in which cells or microorganisms are mixed by mechanically-driven impellers.

stock plant The source plant from which cuttings or **explants** are obtained. Stock plants should be well maintained to optimize **explant**

and cutting quality.

stock solution Pre-prepared solution of commonly used reagents.

stock The lower portion of a graft. *See*: **rootstock.**

stolon A lateral **stem** that grows horizontally along the ground surface. Used by some plant species as a mechanism for dispersal, since stolon nodes can differentiate into normal stems and roots, giving rise to a daughter plant removed from the parent.

stoma (pl.: stomata) 1. Any of various small openings or pores in an animal body, especially an opening resembling a mouth in various invertebrates. 2. A pore in the **epidermis** of the leaf or **stem** of a plant, which allows the exchange of gases, including water vapour, to and from the intercellular spaces. Sometimes used loosely to refer to the pore along with its associated pair of guard cells. *Synonym*: stomate. *See*: **stomatal complex.**

stomatal complex Includes the **stoma**, together with its guard cells and, when present, any related subsidiary cells.

stomatal index A measurement of the surface density of **stomata**. This **parameter** has been found useful in comparing leaves of different sizes. Relative humidity and light intensity during leaf development affect the value of stomatal index.

stop codon A set of three **nucleotides** for which there is no corresponding **tRNA** molecule to insert an **amino acid** into the **polypeptide** chain. **Protein** synthesis is hence terminated and the completed polypeptide released from the **ribosome.** Three stop codons are known: UAA (ochre), UAG (amber) and UGA (opal). *Synonyms*: **chain terminator**; nonsense codon, **termination codon.**

STR Abbreviation for **sequence tandem repeat**. *See*: **tandem repeat.**

strain A group of individuals derived by descent from a single individual within a species.

stratification Subjection of moist seeds to a period of low temperature (+2 to +4 °C) to break

dormancy.

streptavidin A microbial **protein** with a high affinity for the B complex **vitamin biotin**. The specific interaction of these two molecules has been exploited in **labelling** technology and in applications where a specific molecule needs to be captured or purified.

stress Non-optimal conditions for growth. Stresses may be imposed by biotic (pathogens, pests) or **abiotic** (environment, such as heat, drought etc.) factors.

stress protein *See*: **heat shock protein**.

stringency Reaction conditions (notably temperature, salt concentration and pH) that affect the annealing process of **single-stranded DNA** or **RNA** to make **double-stranded DNA** or RNA, or **DNA/RNA** hybrids. At high stringency, duplexes form only between strands with perfect **complementarity**; lower stringency allows the annealing of strands with some degree of **mismatch**.

stringent plasmid A **plasmid** that can only replicate at the same time as does the main bacterial chromosome, and is present as a single or, at most, several copies per cell.

stroma The supporting connective **tissue** of an **organ** or **plastid**.

structural gene A gene that encodes a polypeptide, with either enzymatic or structural functions, and that is required for the normal **metabolism** and growth of a cell or organism.

structure-functionalism The scientific tradition that stresses the relationship between a physical structure and its function, e.g. the related disciplines of anatomy and physiology.

STS Abbreviation for **sequence-tagged site**.

style Slender column of tissue that arises from the top of the **ovary** and terminates in the **stigma**, and through which the **pollen** tube must grow to achieve fertilization.

sub-clone A procedure in which a large cloned **DNA** molecule is divided into smaller fragments, each one of which is then separately cloned.

sub-culture Division and transfer of a portion

of a culture to fresh medium. Sometimes used to denote the adding of fresh liquid to a suspension culture. *Synonym*: passage.

sub-culture interval The time between consecutive sub-cultures of cells.

sub-culture number The number of times cells, etc., have been sub-cultured.

subgenomic promoter A **promoter** added to a **virus** for a specific **heterologous** gene, resulting in the formation of **mRNA** for that **gene** alone.

subspecies Population(s) of organisms sharing certain characteristics that are not present in other populations of the same species.

sub-strain Derived from a **strain** by the isolation of an individual or group of individuals having properties or markers not shared by the strain as a whole.

substrate 1. A compound that is altered by an enzyme. 2. Food source for growing cells or microorganisms. 3. Material on which a sedentary organism lives and grows.

sub-unit vaccine One or more immunogenic **proteins**, either purified from the **pathogen** itself or produced from a cloned pathogen gene. A **vaccine** composed of a purified **antigenic** determinant that is separated from the virulent organism.

sucker A shoot that arises from an underground **root** or **stem**. Of particular significance to grafted plants, since the sucker will be genotypically **rootstock**, rather than **scion**.

suckering Type of vegetative **propagation** where lateral buds grow out to produce an individual that is a **clone** of the parent.

sucrose density gradient centrifugation A procedure used to fractionate nucleic acids on the basis of their size.

superbug Jargon for a particular engineered **strain** of *Pseudomonas*, in which various hydrocarbon-degrading genes, derived from different plasmids, were combined into one genotype. This provided the basis for the precedent-setting legal decision that declared that genetically engineered organisms were patentable. *See:* **Chakrabarty decision**.

supercoil The conformation of a **double-stranded DNA** molecule placed under torsional **stress** as a result of interactions with **proteins**. The stress is accommodated by a twist imposed on the duplex. A left-handed supercoil favours unwinding of **the double helix**; a right-handed supercoil favours tighter winding.

supercoiled plasmid The predominant *in vivo* form of most **plasmids**, in which the **DNA** is coiled around **histone**-like **proteins**. When supporting **proteins** are stripped away during **DNA** extraction from the bacterial cell, the plasmid molecule also tends to supercoil around itself *in vitro*.

supergene A group of tightly linked genes that are co-inherited, and may be functionally related.

supernatant The liquid phase remaining after insoluble materials are pelleted by **centrifugation** or precipitation.

suppressor mutation A mutation that reverses the effect of an earlier mutation, e.g. a mutation in a **gene** for a **tRNA** that permits it to read and override an amber mutation.

suppressor-sensitive mutant An organism that can grow in the presence, but not in the absence of a second genetic factor (the suppressor).

susceptible Inability to withstand injury due to biotic or **abiotic stress**. *Opposite*: **resistance, tolerance.**

suspension culture A type of culture in which cells and/or clumps of cells grow and multiply while suspended in a liquid medium.

symbiont An organism living in **symbiosis** with another, dissimilar organism.

symbiosis The close association of two different kinds of living organisms where there is benefit to both or where both receive an advantage from the association. A prominent example is the colonization of ***Rhizobium*** spp. inside the roots of leguminous plants.

sympatric speciation The evolution of new **species** by **populations** that inhabit the same or overlapping geographic regions.

sympodial A type of plant development in which the **terminal bud** of the **stem** stops growing due either to its abortion, or to its differentiation into a floral **meristem**. Frequently, the uppermost **lateral bud** then takes over the further axial growth of the stem.

synapsis Synonym of chromosome pairing.

synaptonemal complex (Abbreviation: SC). A ribbon-like **protein**aceous structure formed between paired **homologous chromosomes** at the end of the first meiotic **prophase**. The SC binds the **chromatids** along their length, and facilitates **crossing over**.

synchronous culture A culture in which the **cell cycle** is synchronized for the majority of the cells present. Synchrony can be induced by the addition of drugs which arrest the cell cycle at specific stages.

syncytium A group of cells in which cytoplasmic continuity is maintained; the effect is of a multinucleate cell.

syndrome A group of specific characters that occur together, and are characteristic of a particular disease or genetic condition (e.g. Down's syndrome).

synergid One of the two **haploid** nuclei at the micropylar end of the **embryo sac** of higher plants. The third **nucleus** is the **egg**.

synergism An interaction between two organisms (e.g. ***Rhizobium*** and legumes) in which the growth of one is helped by the other. *Opposite*: **antagonism**.

syngamy Synonym of **fertilization**.

synkaryon The initial **hybrid nucleus** of the **zygote**, formed by the fusion of the gametic nuclei upon fertilization. A hybrid nucleus formed by the fusion of two different somatic cells during **somatic cell hybridization** is called a **heterokaryon**.

synteny The occurrence of two or more loci on the same **chromosome**, without regard to their genetic linkage. Increasingly used to describe the **conservation** of **gene** order between related **species**.

T

T Abbreviation for **thymine**.

T cell Lymphocytes which pass through the thymus **gland** during maturation. Different kinds of T cells play important roles in the **immune response**. *Synonym*: T lymphocyte. *See*: **T-cell-mediated (cellular) immune response**.

T cell receptor An **antigen**-binding **protein**, located on the surface of mammalian killer T cells, which mediates the cellular immune response. **T cell antigen** encoding genes are assembled from gene segments by **somatic recombination** processes that occur during **lymphocyte** differentiation.

T lymphocyte *See*: **T cell**.

T_0, T_1 and T_2 Successive generations of plants following a **transformation** event. The parent transformed plant is T_0, its immediate **progeny** is T_1, and the progeny of the T_1 are T_2 plants etc. Of particular interest is the stability of **transgene** expression from T_0 to T_2, and beyond.

T4 DNA ligase An enzyme, present in bacteria infected with **bacteriophage** T4, which catalyses the joining (**ligation**) of, and repairs nicks in, **duplex DNA** molecules. Ligation activity requires that one **DNA** molecule has a 5'-phosphate group, and that the other has a free 3'-hydroxyl group.

tag *See*: **label**.

tailing The *in vitro* addition, to the 3'-hydroxyl ends of a **double-stranded DNA** molecule, of multiple copies of a single **nucleotide** by the enzyme **terminal transferase**. *Synonym*: **homopolymeric tailing.**

tandem array *See*: **tandem repeat.**

tandem repeat Two (or more) contiguous identical **DNA** sequences. The orientation can be either head-to-tail, or head-to-head. *Synonyms*: tandem array, sequence tandem repeat.

tank bioreactor A fermentation vessel designed to grow large scale quantities of a **microorganism** (bacteria, **yeast** or fungi). Most tank **bioreactors** are designed to be stirred mechanically, since this allows effective distribution throughout the culture of gas and nutrients. Alternative bioreactors use fibre or membrane surfaces to immobilize the cultured cells.

taproot Root system in which the **primary** root has a much larger diameter than any lateral roots (e.g. carrot). *Opposite*: **fibrous root**.

***Taq* polymerase** A heat-stable **DNA polymerase** isolated from the thermophilic bacterium *Thermus aquaticus,* widely used in **PCR**.

target In diagnostic tests, the molecule or **nucleic acid** sequence assayed in a sample. In **mutagenesis**, the **gene sequence** that needs to be altered to generate the desired change in **phenotype.**

target site duplication A short sequence of **DNA** duplicated when a **transposable element** inserts at a new locus; usually found at each end of the insertion.

targeted drug delivery A method of delivering the activated form of a **drug** molecule to the site in the body where it is needed, rather than allowing it reach the target by uncontrolled diffusion.

targeting vector A cloning **vector** carrying a **DNA** sequence capable of participating in a recombinational event at a specified chromosomal location in the **host** cell.

TATA box A widely conserved adenine- and thymine-rich **DNA** sequence found 25-30 bp **upstream** of the **transcription** initiation point of many eukaryotic genes. The TATA box is implicated in the promotion of gene **transcription** as it acts as a binding site for **RNA polymerase**. Analogous to the **Pribnow box** in prokaryotic promoters. *Synonym*: Hogness box.

tautomeric shift The transfer of a hydrogen atom from one position in an organic molecule to another position. Tautomers can have widely different biological activities, as the shift can induce a significant change in the **conformation** of the **molecule**.

tautomerism A type of isomerism in which the two isomers arising from a **tautomeric shift** are in equilibrium.

T-cell-mediated (cellular) immune response The synthesis of antigen-specific **T cell** receptors and the development of killer T cells in response to an encounter of immune system cells with an unrecognized immunogenic molecule.

T-DNA The **DNA** segment of the **Ti plasmid**, present in pathogenic ***Agrobacterium tumefaciens***, that is transferred to plant cells and inserted into the plant's **DNA** as part of the **infection** process. **Wild type** T-**DNA** encodes enzymes that induce the plant to synthesize specific **opines** that are required for bacterial growth. In engineered T-**DNAs**, these genes are replaced by a **transgene**(s).

telomerase An enzyme that maintains the structure of the **telomere** by adding the required repetitive sequences to the ends of eukaryotic chromosomes.

telomere The structure found at the end of eukaryotic chromosomes containing specialized repetitive (and widely conserved across species) **DNA** sequences, which are necessary to assure the completion of a cycle of **DNA** replication.

telophase The last stage in each mitotic or meiotic division, in which the chromosomes coalesce at each pole of the dividing cell.

temperate phage A phage (**virus**) that invades but does not normally destroy (lyse) the host bacterial cell. Under specific circumstances, the **lytic cycle** is induced, resulting in the release of infective phage particles.

temperature-sensitive mutant An organism that can grow at one temperature but not at another.

temperature-sensitive protein A **protein** that is functional at one temperature but loses function at another (usually higher) temperature.

template An **RNA** or **single-stranded DNA** molecule, used by **polymerases** to generate a **complementary nucleotide** strand.

template strand *See*: **anticoding strand**.

terminal bud A branch tip, an undeveloped shoot containing rudimentary floral buds or leaves, enclosed within protective bud scales.

terminal transferase An enzyme that catalyses the addition of **nucleotides** to the 3' end of a **DNA** molecule.

terminalization Repelling movement of the centromeres of bivalents in the **diplotene** stage of the meiotic prophase, that appears to move visible chiasmata toward the ends of the bivalents.

termination codon *See*: **stop codon**.

termination signal In **transcription**, **a nucleotide** sequence that specifies **RNA** chain termination.

terminator 1. A **DNA** sequence just **downstream** of the **coding** segment of a gene, which is recognized by **RNA polymerase** as a signal to stop synthesizing **mRNA**. 2. A term used in **GMO** technology for a **transgenic** method which genetically sterilizes the progeny of the planted **seed**, thereby preventing the use of farm-saved seed.

terminator codon *See*: **stop codon**.

terminator gene A specific variety-level **genetic use restriction technology**. A patented technique.

terminator region A **DNA** sequence that signals the end of **transcription**.

tertiary structure The three-dimensional **conformation** taken up by complete macromolecules as a result of intramolecular interactions, such as hydrogen-bonding. *See*: **primary structure**, **secondary structure**, **quaternary structure**.

testcross A cross between a genetically unknown individual and a **recessive** tester to determine whether the individual in question is **heterozygous** or **homozygous** for a certain **allele**. It can also used as a method to test for **linkage**, i.e. to estimate **recombination** fraction.

testis (pl.: testes) Male sex **organ** where **spermatozoa** mature and are stored.

testosterone Male hormone, synthesized in the **testis** of mammals; used to induce sex reversal in fish.

test-tube fertilization *See*: ***in vitro* fertilization**.

tetracycline An **antibiotic** that interferes with **protein synthesis** in prokaryotes. A **gene** encoding **resistance** to tetracycline has been widely used as a **marker** to distinguish between transformed and non-transformed cells in the production of **transgenic** plants.

tetrad The four **haploid** cells formed after the second meiotic division in plants (pollen tetrads) or fungi (ascospores).

tetraploid An organism, or a tissue whose cells contain four **haploid** sets of **chromosomes**.

tetrasomic (*Noun*: tetrasome). Pertaining to a nucleus or an organism with four members of one of its chromosomes, whereas the remainder of its **chromosome** complement is diploid. Chromosome formula: 2n + 2.

tetratype In fungi, a **tetrad** of spores that contains four different types; e.g. *AB*, *aB*, *Ab* and *ab*.

TGGE Abbreviation for **thermal gel gradient electrophoresis**.

thallus Plant body without true roots, stems, or leaves.

therapeutic agent A compound used for the treatment of a disease or for improving the well-being of an organism. *Synonyms*: pharmaceutical agent, **drug**.

therapeutic cloning The potential use of **stem cells** to grow, *in vitro*, tissue or organs for use in transplantation. Because these cells would be obtained from, and would therefore be genetically identical to the patient's own cells, problems of transplant rejection would be overcome. The technique would also remove the difficulty of identifying an organ donor.

thermal gel gradient electrophoresis (Abbreviation: TGGE). A method for separating **DNA** fragments according to their mobility under increasingly denaturing conditions imposed by heat.

thermal shock Exposure to reduced or increased temperature for a significant period.

thermolabile Not resistant to heat, often in the context of a molecule which is unstable upon heating. *Opposite*: thermostable.

thermophile An organism which is adapted to high temperatures, such as in hot springs and geysers, smoker vents on the sea floor, and domestic hot water pipes. A wide range of bacteria, fungi and simple plants and animals can grow at temperature up to 50 °C; thermophiles grow and reproduce at above 50 °C. They can be classified, according to their optimal growth temperature, into simple thermophiles (50-65 °C); thermophiles (65-85 °C), and extreme thermophiles (>85 °C). *See*: **mesophile, psychrophile.**

thermosensitivity Loss of biological activity of a **molecule** at high temperature.

thermostable A molecule which retains its biological activity at some specified higher temperature. *Opposite*: thermolabile.

thermotherapy Exposure to elevated temperatures, a technique mainly used for **virus** or mycoplasma elimination, taking advantage of the higher thermostability of the **host** over its **pathogen**. *Synonym*: **heat therapy**.

thinning 1. Removal of older stems to promote new growth. 2. Removal of excess fruits to improve the size and quality of the remaining fruits. 3. Removal of seedlings spaced too closely for optimum growth.

thymidine kinase (Abbreviation: tk). An enzyme that allows a cell to utilize an alternate metabolic pathway for incorporating **thymidine** into **DNA**. Used as a **selectable marker** to identify transfected eukaryotic cells.

thymidine The deoxyribonucleoside resulting from the combination of the **base thymine** (T) and the sugar 2-deoxy-D-**ribose**. *See*: **TTP**.

thymidine triphosphate Abbreviation: TTP; dTTP is strictly correct but rarely used.

thymidylic acid (Abbreviation: TMP or dTMP). Synonym for **thymidine** 5'-monophosphate, a **deoxyribonucleotide** containing the **nucleoside** thymidine.

thymine (Abbreviation: T). One the **bases** found in **DNA**. *See*: **thymidine**.

Ti plasmid Tumour-inducing **plasmid**. A large plasmid present in pathogenic ***Agrobacterium tumefaciens***, responsible for the **induction** of tumours in plant with **crown gall disease**. Engineered forms of this **plasmid** are central to the production of **transgenics** in many crop species. *See*: **T-DNA**.

tion will absorb water and display swelling and turgidity. *Opposite*: **hypertonic**.

tissue A group of cells of similar structure which sometimes performs a special function.

tissue culture The *in vitro* culture of **cells**, **tissues** or **organs** in a nutrient **medium** under **sterile** conditions.

titre 1. The concentration of infectious **virus** particles present in a suspension. 2. A measure of **antibody** concentration, given by the highest dilution of the sample that results either in a useable **immunoassay**, or in the formation of visible precipitate when challenged by the appropriate antigen.

tk Abbreviation for **thymidine kinase**.

TMP Abbreviation for the deoxyribonucleotide **thymidine** 5'-monophosphate. *See*: **thymidylic acid**.

tolerance Incomplete **resistance** to a given biotic or **abiotic stress**. Tolerant genotypes are less inhibited by the stress, but are not immune.

tonoplast The cytoplasmic membrane bordering the **vacuole** of plant cells. It plays a prominent role in regulating the **osmotic** pressure exerted by the cell sap.

topo-isomerase *See*: **DNA topo-isomerase.**

totipotency The ability of a **cell** or **tissue** to be induced to regenerate into a complete **organism.**

totipotent (adj.) *See*: **totipotency.**

toxicity The extent to which a toxic compound negatively affects a given trait.

toxin A compound produced by one **organism,** which is deleterious to the growth and/or survival of another organism of the same or different species.

tracer A substance (typically a radioactive **isotope** or a fluorescent dye) that can be detected by physical means, and which is used to analyse the progress of a chemical reaction or a biological process.

tracheid An elongated, tapering **xylem** cell, with lignified pitted walls, adapted for solute conduction and physical support. Found in conifers, ferns and related plants.

trait One of the many characteristics that define an organism. The **phenotype** is a description of one or more traits. *Synonym:* character.

trans **configuration** *See*: **repulsion.**

trans **heterozygote** A double **heterozygote** that contains two mutations arranged in the *trans* configuration.

trans **test** *See*: **complementation test.**

***trans*-acting** 1. A term describing substances that are diffusable and that can affect spatially separated entities within cells. 2. A genetic element (e.g. a **promoter** sequence) that is effective only when present in the *trans* configuration.

***trans*-acting factor** Any of the multiple ancillary **DNA**-binding **proteins** that interact with the *cis*-regulatory **DNA** sequences to control gene expression.

transcapsidation The partial or full coating of the **nucleic acid** of a **virus** particle with the **coat protein** of a different virus.

transcript An **RNA** molecule that has been synthesized from a specific **DNA template**. In eukaryotes, the **primary** transcript produced by **RNA polymerase** is often processed or modified in order to form functional **mRNA**, **rRNA** or **tRNA**. *See*: **splicing**.

transcription factor A **protein** that regulates the **transcription** of genes.

transcription Synthesis of **RNA** from a **DNA template** via **RNA polymerase**.

transcription unit A segment of **DNA** that contains signals for the **initiation** and termination of **transcription**, and is transcribed into one **RNA** molecule.

transcriptional anti-terminator A **protein** that prevents **RNA polymerase** from terminating **transcription** at specific transcription termination sequences.

transcriptional roadblock A DNA-binding **protein** which affects the rate at which **RNA polymerases** transcribe genes. The **protein/DNA** complex interferes with the passage of the elongation complex. In some cases, these obstacles are readily bypassed, but in others, a significant level of pausing or termination occurs, and this can then act as a control point for **gene** expression.

transducing phage *See*: **transduction**.

transduction 1. Genetic: the transfer by means of a viral **vector** of a **DNA** sequence from one cell to another. 2. Signal: any process that helps to produce biological responses to events in the environment (e.g. transduction of **hormone** binding into cellular events by **hormone** receptors).

transfection The **infection** of a **cell** with isolated viral **DNA** (or **RNA**), resulting in the production of intact viral particles.

transfer RNA *See*: **tRNA**.

transferase A class of enzymes that catalyses the transfer of a group of atoms from one **molecule** to another.

transformant A cell or organism that has been genetically altered through the integration of a **transgene**(s). Primary: the first generation following the **transformation** event. Secondary: progeny of the primary transformant.

transformation 1. The uptake and integration of **DNA** in a cell, in which the introduced **DNA** is intended to change the **phenotype** of the recipient **organism** in a predictable manner. 2. The conversion, by various means, of cultured animal cells from controlled to uncontrolled cell growth, typically through **infection** with a **tumour virus** or **transfection** with an **oncogene**.

transformation efficiency or frequency The fraction of a cell population that takes up and integrates the introduced **transgene**; expressed as the number of transformed cells recovered divided by the total number of cells in a **population**.

transforming oncogene A gene that, upon **transfection**, converts a previously immortalized **cell** to the malignant **phenotype**.

transgene An isolated **gene sequence** used to transform an **organism**. Often, but not always, the transgene has been derived from a different species than that of the recipient.

transgenesis The introduction of a **gene** or genes into animal or plant cells, which leads to the transmission of the input gene (**transgene**) to successive generations.

transgenic An individual in which a **transgene** has been integrated into its **genome**. In transgenic eukaryotes, the transgene must be transmitted through **meiosis** to allow its **inheritance** by the **offspring.**

transgressive variation The appearance, in a segregating generation, of individuals showing expression of a **trait** outside the extremes defined by the parent of the **cross** that was used to generate the **population.**

transient expression Short-term activity of a **transgene** following its introduction into **target tissue**. Transient expression usually implies non-integration of the transgene into the **host genome**.

transition stage The period between juvenile and reproductive stages of growth.

transition The substitution in **DNA** or **RNA** of one purine by another **purine**, or of one **pyrimidine** by another pyrimidine. *See*: **transver-**

sion, base substitution.

transition-state intermediate In a chemical reaction, an unstable and high-energy configuration assumed by reactants on the way to making products. **Enzymes** are thought to bind and stabilize the transition state, thus lowering the energy of activation needed to drive the reaction to completion.

translation The process of **polypeptide** synthesis in which the **amino acid sequence** is determined by **mRNA**, mediated by **tRNA** molecules, and carried out on **ribosomes**.

translational initiation signal *See*: **initiation codon**.

translational start codon *See*: **initiation codon**.

translational stop signal *See*: **termination codon**.

translocation 1. The movement of nutrients or products of **metabolism** from one location to another. 2. Change in position of a segment of a **chromosome** to another, non-homologous chromosome.

transposable (genetic) element A **DNA** element that can move from one location in the **genome** to another. *Synonym*: **transposon**.

transposase An enzyme encoded by a **transposon gene** that catalyses the movement of a **DNA** sequence to a different site in a **DNA** molecule.

transposition The process whereby a **transposon** or **insertion sequence** inserts itself into a new site on the same or another **DNA** molecule. The exact mechanism is not fully understood and different transposons may transpose by different mechanisms. Transposition in bacteria does not require extensive **DNA homology** between the transposon and the **target DNA.**

transposon Synonym of **transposable genetic element**.

transposon tagging A method of gene isolation that exploits the disruption of normal **gene expression** that is the result of an insertion of a **transposon** within, or close to the **target**. Since the sequence of the transposon is known, this can be used as a **DNA probe** to define the **DNA** fragment containing the target gene. Large-scale experiments to generate populations of gene **mutations** are colloquially referred to as **gene machines.**

transversion The substitution in **DNA** or **RNA** of one **purine** by a **pyrimidine** or *vice versa. See*: **transition, base substitution.**

tribrid protein A fusion **protein** that has three segments, each encoded by parts of different genes.

trichome A short filament of cells, resulting in a hair-like structure.

tri-hybrid The hybrid **offspring** of a **cross** between parents carrying contrasting **alleles** at three loci.

trinucleotide repeat Tandem repeats of three **nucleotides** that are present in many genes. Commonly trinucleotide repeats have undergone variable expansion in copy number, forming the basis of **microsatellite** markers, and occasionally resulting in the formation of alleles giving rise to genetic disease.

tripartite mating A process in which **conjugation** is used to transfer a **plasmid vector** to a **target cell** when the plasmid vector is not self-mobilizable.

triplet A sequential group of three **nucleotides** in **DNA** or **RNA**. *See*: **codon.**

triploid A cell, tissue or organism containing three times the **haploid** number of **chromosomes.**

trisomic (adj.) *See*: **trisomy.**

trisomy The presence in a diploid **cell** or **organism** of an extra **chromosome** of one **homologue** (chromosome formula: 2n + 1). *See*: **disomy; monosomic.**

triticale The hybrid man-made **species** formed by the crossing of **tetraploid** or hexaploid wheat with **diploid** rye.

tRNA Abbreviation for transfer **RNA**. Small **RNA** molecules that transfer **amino acids** to the **ribosome** during **protein** synthesis. Each **tRNA** binds just one species of **amino acid** and recognizes a specific **codon** in the m**RNA**, thus implementing the **genetic code.**

tropism Plant response to an external stimulus,

resulting in the bending/turning of **stem** or **root** growth. Typical tropisms are **phototropism** (light), **geotropism** (gravity) or hydrotropism (water).

true-to-type Conforming to the **phenotype** of the breed/variety.

trypsin A **proteolytic** enzyme used *in vivo* for the digestion of **peptides**. It acts by hydrolysing peptide bonds on the carboxyl side of the amino acids arginine and lysine.

trypsin inhibitor Substances inactivating **trypsin**, typically found in **seed tissue** of certain plants, where they are thought to have evolved as anti-feedant agents against insect predators.

TTP Abbreviation for **thymidine** 5'-triphosphate. TTP is required for **DNA** synthesis since it is a direct precursor molecule. *See*: **thymidine, thymidylic acid**.

tubulin The major **protein** component of the **microtubules** of eukaryotic cells.

tumble tube A glass tube mainly used ***in vitro*** to agitate and consequently aerate suspension cultures. The tube, which is commonly attached to a slowly revolving platform, is closed at both ends, with a side-neck opening.

tumor-suppressor gene A gene that regulates cell growth. If such a gene becomes dysfunctional, and potentiating damage occurs to the **cell**, then uncontrolled growth and a cancer may result. *See*: **p53 gene, oncogene**.

tumour virus A virus capable of transforming a cell to a malignant **phenotype**.

tumour-inducing plasmid *See*: **Ti plasmid**.

tunica The outer one- to four-cell layer region of the apical meristem, where **cell division** is anticlinal, i.e. perpendicular to the surface. *See*: **apical meristem**.

turbidostat An open **continuous culture** in which a pre-selected **biomass** density is uniformly maintained by automatic removal of excess cells. The fresh medium flows in response to an increase in the turbidity (usually corresponding to cell density) of the culture.

turgid Swollen, distended; referring to a **cell**

that is extended as a result of adequate water uptake. Loss of turgidity in plant cells is a sign of water deficit.

turgor potential *See*: **pressure potential**.

turgor pressure The pressure within a **cell** resulting from the absorption of water into the **vacuole** and the **imbibition** of water by the protoplasm.

turion An underground **bud** or shoot from which an aerial **stem** arises. *See*: **sucker**.

twin One of two individuals originating from the same **zygote**.

U

U Abbreviation for **uracil**.

ubiquitin A small **protein**, present in all eukaryotic cells, which plays an important role in tagging **protein**s destined for **proteolytic** cleavage (because they are damaged or no longer needed).

ultrasonication *See*: **sonication**.

UMP Abbreviation for the (ribo)**nucleotide uridine** 5'-monophosphate. *See*: **uridylic acid**.

understock Host plant for a grafted scion, a branch or shoot from another plant; an understock may be a fully grown tree or a stump with a living **root** system.

undifferentiated Undifferentiated cells are those which have not been committed to become part of a specialized tissue.

unencapsidated A virus not enclosed by a **coat protein** or **capsid**.

unequal crossing over Abnormal meiotic event, in which one **chromatid** contains a **duplication** and the other a **deletion**. Often arises in a region containing repeated **DNA** sequences, which can pair out of register.

unicellular Tissues, organs or organisms consisting of a single cell.

uniparental inheritance The inheritance of genes exclusively from one parent, e.g. **chloroplast DNA** is inherited either maternally (many angiosperms) or paternally (most gymnosperms).

unisexual Higher organisms (animals or plants) possessing either male or female reproductive organs, but not both.

univalent An unpaired chromosome at the first division of meiosis.

universal donor cell Cells that, after introduction into a recipient, will not induce an **immune response** that leads to their rejection.

universality Referring to the genetic code, the **triplet codons** are translated to the same **amino acid**, with minor exceptions, in virtually all species.

unorganized growth *In vitro* formation of tissues with few differentiated **cell** types and no recognizable structure; typical structure of calli formed in **tissue culture**. *Opposite*: **organized growth**.

upstream 1. The stretch of **DNA** lying in the 5' direction from the site being considered. Where the reference point is the **initiation** of transcription, the first transcribed base is designated +1 and upstream nucleotides are marked with minus signs, e.g. -1, -10; 2. In chemical engineering, those phases of a manufacturing process that precede the **biotransformation** step. Refers to the preparation of raw materials for a **fermentation** process. Also called upstream processing.

upstream processing *See*: **upstream** (2).

uracil (Abbreviation: U). One the **bases** found in **RNA**. *See*: **uridine**.

uridine The (ribo)**nucleoside** resulting from the combination of the **base uracil** (U) and the sugar D-**ribose**. *See*: **uridylic acid**, **uridine triphosphate**.

uridine triphosphate (uridine 5'-triphosphate) (Abbreviation: UTP). Required for **RNA** synthesis since it is a direct precursor molecule. *See*: **uridylic acid**.

uridylic acid Synonym for **uridine** 5'-monophosphate (abbreviation: UMP), a (ribo)**nucleotide** containing the **base** uracil. *See*: **uridine triphosphate**.

utilization of farm animal genetic resources The use and development of animal genetic resources for the production of food in a sustainable system of agriculture.

UTP *See*: **uridine triphosphate**.

V region Variable region in antibodies. *See*: **CDR**.

v/v Abbreviation for volume per volume. The relative proportion of each liquid in a mixture.

vaccination *See*: **preventive immunization**.

vaccine A preparation of dead or attenuated (weakened) **pathogens**, or of derived antigenic determinants, that can induce the formation of antibodies in a **host**, and thereby produce host **immunity** against the pathogen. *See*: **sub-unit vaccine, viral vaccine, DNA vaccine, inoculum**.

Vaccinia The cowpox **virus** used to vaccinate against smallpox and, experimentally, as a **carrier** of genes for antigenic determinants cloned from other disease organisms.

vacuole A fluid-filled membrane-bound cavity inside many plant cells, in which various plant products and by-products are stored.

variable domain Regions of **antibody** molecules that have different **amino acid** sequences in different antibody molecules. These regions are responsible for the antigen-binding **specificity** of the antibody.

variable expressivity Variation in the **phenotype** caused by different **alleles** of the same **gene** and/or by the action of other genes and/or by the action of non-genetic factors.

variable number tandem repeat (Abbreviation: VNTR). A **DNA** sequence, present as tandem repeats, for which the number of copies varies greatly between unrelated **genotypes**.

variable surface glycoprotein (Abbreviation: VSG). One of a battery of **antigenic determinants** expressed by a **microorganism** to elude immune detection.

variance A statistical term representing a measure of the dispersion of data from the overall mean. Used to quantify the variability of a population.

variant An individual that is genetically distinct from others in the population.

variation Differences between individuals within a **population** or among populations.

variegation The occurrence, within a single tissue, organ or organism, of mosaicism. Usually referring to plants showing either both green and **albino** colouration in a leaf, or flecks of contrasting colour in a flower. The origin of variegation can be through viral **infection**, nutritional deficiency, or genetic **instability** caused by **transposon** activity. *See also:* **chimera**

variety 1. A naturally occurring subdivision of a **species**, with distinct morphological characters. 2. A defined **strain** of a crop plant, selected on the basis of phenotypic (sometimes genotypic) homogeneity.

vascular bundle A strand of tissue containing **primary xylem** and primary **phloem** (and **procambium** if present) and frequently enclosed by a bundle sheath of **parenchyma** or fibres.

vascular cambium In biennials and perennials, cambium giving rise to **secondary phloem** and secondary **xylem**.

vascular plant Plant species possessing organized **vascular** tissues.

vascular Plant **tissue** specialized for the conduction of water or nutrients

vascular system 1. A specialized network of vessels for the circulation of fluids throughout the body tissue of an animal. 2. The system of **vascular tissue** in plants.

vector 1. An organism, usually an insect, that carries and transmits **pathogens**. 2. A small **DNA molecule** (plasmid, virus, bacteriophage, artificial or **cut DNA** molecule) that can be used to deliver **DNA** into a cell. Vectors must be capable of being replicated and contain **cloning** sites for the introduction of foreign **DNA**.

vegetative propagation *See*: **asexual propagation.**

velocity density gradient centrifugation A procedure used to separate **macromolecules** based on their rate of movement through a density gradient.

velogenetics The combined use of **marker-assisted selection** and **embryo** technologies such as **OPU, IVM** and **IVF**, in order to increase the rate of genetic improvement in animal populations.

vermiculite Material made from expanded mica used as a rooting medium and as a soil additive.

vernalization Chilling juvenile plants for a minimum period in order to induce flowering. Some plants require vernalization to flower, but others have no such requirement.

vessel A series of **xylem** elements whose function is to conduct water and nutrients in plants.

vessel element A type of **cell** occurring within the **xylem** of flowering plants. Many are water-conducting vessels.

viability test Assay of the number or percent of living cells or plants in a **population** following a specific treatment. Often used to describe quality of **seed** following long-term storage.

viability The capability to live and develop normally.

viable Capable of normal completion of life cycle.

vibrio Comma-shaped **bacterium**.

vir **genes** A set of genes on a **Ti plasmid** that prepare the **T-DNA** segment for transfer into a plant cell.

viral coat protein A **protein** present in the layer surrounding the **nucleic acid** core of a **virus**.

viral oncogene A viral gene that promotes tumour development in a **host**.

viral pathogen A disease-causing virus.

viral vaccine Vaccine consisting of live viruses, genetically engineered to avoid causing the disease itself.

virion A complete infectious **virus** particle.

viroid A plant pathogenic agent, composed of an infectious single-stranded low molecular weight **RNA**, and no **coat protein**.

virulence The degree of ability of an **organism** to cause disease. The relative infectiousness of a bacterium or **virus**, or its ability to overcome the resistance of the **host** metabolism.

virulent phage A **phage** that destroys its **host** bacterium.

viruliferous A vector (usually insect) organism that carries virions and spreads the **virus** from **host** to host by mechanical means.

virus An infectious particle composed of a **protein capsule** and a **nucleic acid** core (**DNA** or **RNA**), which is dependent on a **host** organism for **replication**.

virus-free Plant, animal, **cell**, **tissue** or **meristem** which exhibits no viral symptoms or contains no identifiable **virus** particles.

virus-tested Description of a **organism** or a **cell** stock certified as being free of certain specified **viruses** following recognized procedures of virus diagnosis.

vitamin Naturally occurring organic substance required by living organisms in small amounts to maintain normal health.

vitrified Cultured **tissue** having leaves and sometimes stems with a glassy, transparent or wet and often swollen appearance. The process of vitrification is a general term for a variety of physiological disorders that lead to **shoot tip** and leaf necrosis. *Synonym*: water soaked.

viviparous (adj.) *See*: **vivipary.**

vivipary 1. A form of reproduction in animals in which the developing **embryo** obtains its nourishment directly from the mother via a placenta or by other means. 2. A form of **asexual reproduction** in certain plants, in which the flower develops into a bud-like structure that forms a new plant when detached from the parent. 3. The development of young plants in the inflorescence of the parent plant.

V_{max} The maximal rate of an enzyme-catalysed reaction. V_{max} is the product of E_o (the total amount of **enzyme**) and K_{cat} (the catalytic rate constant).

VNTR Abbreviation for **variable number tandem repeat.**

volatilization The conversion of a solid or liquid into a gas or vapour.

VSG Abbreviation for **variable surface glycoprotein.**

W

w/v Abbreviation for weight per volume. The relative proportions of solid and liquid in a solution.

walking *See*: **chromosome walking**; **primer walking**.

wall pressure Pressure that a **cell wall** exerts against the turgor of the cell contents. Wall pressure is equal and opposite to the turgor potential.

wash-out The loss of the slower growing **microorganism** when two organisms are being grown together.

water potential The pressure gradient that induces the flow of water, particularly with reference to plant water uptake from the soil, comprising the net effects of suction, solutes and matric forces.

water soaked *See*: **vitrified**.

water stress Occurs when plants are unable to absorb enough water to replace that lost by transpiration. Short-term water stress leads to turgor loss (wilting). Prolonged stress leads to cessation of growth, and eventually plant death.

wax Water-insoluble esters of long-chain acids with long-chain alcohols. Waxes form protective waterproof layers on leaves, stems, fruits, animal fur and integuments of insects.

weed A plant growing where it is not wanted. Generally used to describe plants which colonize readily, and can compete for resources with a planted crop.

weediness The ability of a plant to colonize a disturbed habitat and compete with cultivated species.

western blot A technique whereby a complex mixture of size-separated **proteins** is fixed to a solid support, and then probed with a labelled **antibody**. Useful, for example, for the measurement of levels of production of a specific **protein** in a particular **tissue** or at particular developmental stage.

wet weight *See*: **fresh weight**.

wetting agent A substance (usually a detergent) that improves the contact of a liquid to a solid surface by reducing its surface tension.

wild type The most frequent **allele** or genotype found in nature, or a specified **organism** against which **mutants** are defined.

wilt Drooping of stems and foliage due to loss of cell turgor. May be caused by water **stress** or by disease.

wilting point The moisture content of soil at which plants start to wilt, but not to the extent that they fail to recover when placed in a humid atmosphere. *See*: **permanent wilting point**.

wobble hypothesis An explanation of how one **tRNA** may recognize more than one **codon**. The first two bases of the **mRNA** codon and anticodon pair properly, but the third **base** in the anticodon has some flexibility that permits it to pair with either the expected base or an alternative.

X

x The basic number of chromosomes in a **polyploid** series, monoploid/**haploid** = x; **diploid** = 2x; **triploid** = 3x; etc.

xanthophyll A yellow oxygen-containing carotenoid, present in chloroplasts.

X-chromosome *See*: **sex chromosome**.

xenia The immediate effect of **pollen** on some characters of the endosperm.

xenobiotic A chemical compound that is not produced by, and often cannot be degraded by, living organisms.

xenogeneic Refers to **organ**s, genetically engineered ("humanized") to decrease the chance of rejection, that have been grown in an animal of another species for potential transplant to humans.

xenotransplantation The transplantation of tissue or organs from one **species** to another species, typically from pigs to humans. **Zoonoses** are an important issue.

xerophyte A plant very resistant to drought, typically adapted to extremely dry environments.

X-linked disease A genetic disease caused by an **allele** at a locus on the **X-chromosome.**

X-linked The presence of a gene on the **X-chromosome.**

xylem A complex **tissue** specialized for the conduction of water and mineral nutrients in solution. Xylem may also function as a supporting tissue, particularly secondary xylem.

Y

YAC Abbreviation for **yeast artificial chromosome**.

Y-chromosome *See*: **sex chromosome**.

yeast A unicellular ascomycete fungus, commonly found as a contaminant in plant tissue culture.

yeast artificial chromosome (Abbreviation: **YAC**). A **vector** which can be propagated in budding **yeast** (*Saccharamyces pombe*), consisting of the minimal elements required for a **chromosome** to replicate, and allowing for the **cloning** of large **DNA** fragments (hundreds of **kilobase pairs**).

yeast episomal vector (Abbreviation: YEp). A cloning plasmid **vector** for the **yeast** *Saccharomyces cerevisiae* maintained as an **extrachromosomal** nuclear **DNA** molecule.

yeast extract A mixture of substances from yeast. *See:* **organic complex**.

Z-DNA A form of **DNA**, in which the **double helix** is wound in a left-hand, instead of a right-hand, manner. **DNA** adopts the Z conformation when purines and pyrimidines alternate on each strand, e.g. 5′CGCGCGCG 3′ or 3′GCGCGCGC5′. *Synonym*: zig-zag **DNA**.

zig-zag DNA *See*: **Z-DNA**.

zinc finger A DNA-binding **protein motif**, characterized by two closely spaced cysteine and two histidine **residues** that serve as **ligands** for a single Zn^{2+} ion. When bound, the structure takes on a **conformation** in which **amino acid** side chains protrude in a way that allows interaction with the **DNA** major groove.

zone of elongation The section of the young **root** or shoot just behind the apical **meristem**, in which the cells are enlarging and elongating rapidly.

zoo blot Hybridization of cloned **DNA** from one species to **DNA** from a range of other organisms to determine the extent to which the cloned **DNA** is evolutionarily conserved.

zoo FISH Fluorescence *in situ* hybridization technique, probing metaphase **chromosomes** of one species with **DNA** from another species. The technique allows inferences to be made regarding the evolutionary relationships between species. See: **Fluorescence *in situ* hybridization**.

zoonosis A disease that is communicable from animals to humans.

zoospore A spore that possesses flagella and is therefore motile.

zygonema Stage of meiotic **prophase** during which chromosome **synapsis** occurs.

zygospore A thick-walled resistant spore developing from a **zygote** resulting from the fusion of **gametes** in the course of isogamy.

zygote The **diploid** cell formed by the fusion of two **haploid gametes** during **fertilization** in eukaryotic organisms with sexual reproduction.

zygotene (adj.) *See*: **zygonema**.

zymogen Inactive enzyme precursor that after **secretion** is chemically altered to the active form of the **enzyme**.

Abbreviations and symbols

The following editorial abbreviations are used in the text.

adj.	adjective
bp	base pair
cf	*confer* - compare
kb	Kilobase [pairs]
pl.	plural
°	degree
′	prime
→	right arrow

See Annex 1 for a list of the prefixes for decimal multiples and submultiples of SI units.
See Annex 2 for the Greek alphabet.

Notes on the structure of the terminology

- All units are expressed as SI units unless otherwise indicated.
- The entries are in simple alphabetical order, with spaces and punctuation marks, such as hyphens, ignored.
- Acronyms *per se* are included in their alphabetical position in the text. Greek characters have been spelled out in full (e.g. alfa instead of a).
- All the words in the definitions that are reported in bold are themselves defined terms (or their plurals).

Annex 1
Prefixes for decimal multiples and submultiples of SI units

Factor	Prefix	Symbol	Factor	Prefix	Symbol
10^{18}	exa	E	10^{-1}	deci	d
10^{15}	peta	P	10^{-2}	centi	c
10^{12}	tera	T	10^{-3}	milli	m
10^{9}	giga	G	10^{-6}	micro	μ
10^{6}	mega	M	10^{-9}	nano	n
10^{3}	kilo	k	10^{-12}	pico	p
10^{2}	hecto	h	10^{-15}	femto	f
10	deca	d	10^{-18}	atto	a

Annex 2
The Greek alphabet

Letter	Uppercase type	Lowercase type	Letter	Uppercase type	Lowercase type
alpha	Α	α	nu	Ν	ν
beta	Β	β	xi	Ξ	ξ
gamma	Γ	γ	omicron	Ο	ο
delta	Δ	δ	pi	Π	π, ϖ
epsilon	Ε	ε	rho	Ρ	ρ
zeta	Ζ	ζ	sigma	Σ	σ
eta	Η	η	tau	Τ	τ
theta	Θ	θ	upsilon	Υ	υ
iota	Ι	ι	phi	Φ	φ
kappa	Κ	κ	chi	Χ	χ
lambda	Λ	λ	psi	Ψ	ψ
mu	Μ	μ	omega	Ω	ω

Annex 3
Codons and amino acids

UUU Phe	UCU Ser	UAU Tyr	UGU Cys
UUC Phe	UCC Ser	UAC Tyr	UGC Cys
UUA Leu	UCA Ser	UAA stop (ochre)	UGA stop (opal)
UUG Leu	UCG Ser	UAG stop (amber)	UGG Trp
CUU Leu	CCU Pro	CAU His	CGU Arg
CUC Leu	CCC Pro	CAC His	CGC Arg
CUA Leu	CCA Pro	CAA Gln	CGA Arg
CUG Leu	CCG Pro	CAG Gln	CGG Arg
AUU Ile	ACU Thr	AAU Asn	AGU Ser
AUC Ile	ACC Thr	AAC Asn	AGC Ser
AUA Ile	ACA Thr	AAA Lys	AGA Arg
AUG Met	ACG Thr	AAG Lys	AGG Arg
GUU Val	GCU Ala	GAU Asp	GGU Gly
GUC Val	GCC Ala	GAC Asp	GGC Gly
GUA Val	GCA Ala	GAA Glu	GGA Gly
GUG Val	GCG Ala	GAG Glu	GGG Gly

Some minor variations have been observed, notably in mitochondria and chloroplasts.

Annex 4
Amino acids, one and three letter codes

Amino acid	Three letter code	One letter code
alanine	ala	A
arginine	arg	R
asparagine	asn	N
aspartic acid	asp	D
asparagine or aspartic acid	asx	B
cysteine	cys	C
glutamic acid	glu	E
glutamine	gln	Q
glutamine or glutamic acid	glx	Z
glycine	gly	G
histidine	his	H
isoleucine	ile	I
leucine	leu	L

Amino acid	Three letter code	One letter code
lysine	lys	K
methionine	met	M
phenylalanine	phe	F
proline	pro	P
serine	ser	S
threonine	thr	T
tryptophan	try	W
tyrosine	tyr	Y
valine	val	V